From Coello
to Inorganic Chemistry
A Lifetime of Reactions

PROFILES IN INORGANIC CHEMISTRY

Series Editor:
John P. Fackler, Texas A & M University, College Station, Texas

Current Volumes in this Series:

From Coello to Inorganic Chemistry: A Lifetime of Reactions
Fred Basolo

A Continuation Order Plan is available for this series. A continuation order will bring delivery of each new volume immediately upon publication. Volumes are billed only upon actual shipment. For further information please contact the publisher.

From Coello
to Inorganic Chemistry
A Lifetime of Reactions

Fred Basolo
Northwestern University
Evanston, Illinois

Kluwer Academic / Plenum Publishers
New York, Boston, Dordrecht, London, Moscow

In Loving Memory of

My Beloved Wife, Mary

and

My Parents, Giovani and Catherina Basolo

SERIES PREFACE

A renaissance in the field of inorganic chemistry began in the middle of the 20th century. In the years following the discoveries of A. Werner and S. M. Jørgensen at the turn of the century, the field was relatively inactive. The publication of Linus Pauling's *Nature of the Chemical Bond* in 1938 and World War II shortly thereafter launched this renaissance. The war effort required an understanding of the chemistry of uranium and the synthetic actinide elements that were essential to the production of the atom bomb. There was also a need for catalysts to produce rayon, nylon, synthetic rubber, and other new materials for the war effort. As a result, many gifted chemists applied their talents to inorganic chemistry. *Profiles in Inorganic Chemistry* explores the roles some of the key contributors played in the renaissance and development of the field.

Some of the early leaders in this reawakening are now deceased. Pioneers included John Bailar at the University of Illinois, W. Conard Fernelius, at Pennsylvania State University, and Harold Booth at Western Reserve University, who with some others, started the important series entitled *Inorganic Syntheses*. Several inorganic chemistry journals were born, as were various monograph series including the *Modern Inorganic Chemistry* series of Kluwer Academic / Plenum Publishers. Geoffrey Wilkinson, who along with E. O. Fischer was the first inorganic chemist since Werner to win the Nobel prize, started his career at Harvard in about 1950 but later that decade moved to the University of London's Imperial College. By then, Ron Nyholm already was building a strong inorganic program at the University of London's University College.

Physical and mathematical concepts including group theory gave inorganic

Series Preface

chemists new tools to understand bonding, structure, and dynamics of inorganic molecules. Fischer, Wilkinson, and their contemporaries opened up a new subfield, organometallic chemistry, out of which many metal-based catalysts were developed. It was soon realized that many inorganic minerals play essential roles as catalysts in living systems. As a result, another subfield, bioinorganic chemistry, was born. The discipline of inorganic chemistry today includes persons of many different walks of life, some creating new materials and catalysts, others studying living systems, many pondering environmental concerns with elements such as tin, mercury, or lead, but all focusing on questions outside the normal scope of organic chemistry.

Organic chemistry has enjoyed a long history as a great science, both in Europe and the United States. During the past 15 years or so, many of the U.S. contributors have produced interesting autobiographies as part of an American Chemical Society series entitled *Profiles in Organic Chemistry*. There is also, however, a need to have students and scientists of inorganic chemistry understand the motivating forces that lead prominent living inorganic chemists to formulate their ideas. I am grateful that Kluwer Academic/Plenum Publishers has undertaken to publish this series. These profiles in inorganic chemistry will portray the interesting and varied personalities of leaders who have contributed significantly to the renaissance of inorganic chemistry.

John P. Fackler, Jr.
College Station, Texas

PREFACE

When approached by Professor John Fackler to write my autobiography, my immediate response was "Why should I write this, for nobody would be the least bit interested in it." John pointed out that several such volumes were written by organic chemists, and the same should be done to commemorate inorganic chemistry and chemists. After thinking about this for several days, I decided to go ahead and try to do it. Now that the text of the book is finished, I can honestly say that I enjoyed writing it.

There are several reasons why I am pleased to have written this book. Some of the reasons are mentioned here, without regard to any order of preference. As I kept writing, I became more and more interested in the many things in my life that I had forgotten. This forced me to think of what happened during my early years. It almost made me feel that I was again living those years of my life. My childhood in the little coal mining village of Coello during the depression of the 1930s taught me a lesson that one needs to experience in order to make use of it. The coal miners and their families were destitute, but they managed to survive this period by assisting one another when the going got tough. They had little or no formal educatizon, but yet they were intelligent. The all maintained their spirit, and looked forward to one day becoming USA citizens. Growing up in this environment has been of enormous help to me.

Another reason for deciding to write my autobiography is to leave it for our children so that they can enjoy reading about the lives of their parents. Furthermore, at age 81 and handicapped, I needed something to do that interested me and

Preface

kept me busy. I continue to come to my office at NU in the AM, have lunch with my faculty colleagues, and then leave in the early PM.

As indicated above, there are several reasons for writing this account of my life. I was fortunate in obtaining a Ph.D. in inorganic chemistry in 1943. This was a time when very little research was being done in the US in inorganic chemistry. It was belived that beginning general chemistry covered inorganic chemistry, and that no further course on it need be offered nor was there any reason for doing research in this area. Think of the enviable time I have had, watching inorganic chemistry grow and reach its present status and importance. I hope that some of the young chemists who read this book will better appreciate the birth of inorganic chemistry in the US, after a gestation period of half a century.

Another thing that I would like to think youngsters may glean from this bok is that scientists, generally, but here chemists, are human beings as are other people. Except for our knowledge of some area of science, we are just people with the same likes and dislikes as any other layperson.

That this autobiography has made it to print is entirely due to Janet H. Goranson, my former secretary for 25 years. She worked hard to improve my poorly written English, and she is a whiz on computers which she used extensively. I was so dependent on her that I felt she should be coauthor, and I thank her for making all of this possible.

Then I thank my former Ph.D. (1964) student, John L. Burmeister, for having read the entire book, making some most valuable suggestions. I selected him knowing that he writes very well, since he had written a few accounts of me for special journal issues that were dedicated to me.

I thank Dianne de Haseth for providing information I needed from our chemistry department archives, and for managing to get all of the inorganic faculty here at the same time for a group photo. I thank Jeffrey S. Goranson for scanning and preparing many of the photographs. I thank Jeannette Watt for making photocopies of items as required.

I also thank staff members from the Northwestern University Archives, American Chemical Society, National Academy of Sciences, Gordon Research Conferences, Chemical Heritage Foundation, and North Atlantic Treaty Organization who sent me important information.

Finally, I thank the many people who answered my request and sent me photos of themselves or others so that they could be included in this book.

Fred Basolo
Northwestern University
Evanston, Illinois
April, 2001

CONTENTS

LIST OF ABBREVIATIONS	xvii
LIST OF FIGURES	xix

CHAPTER 1
FROM COELLO TO INORGANIC COORDINATION CHEMISTRY	1
My Parents	1
The Early Years	3
Coello Elementary School	5
Christopher Community High School (CCHS)	7
Southern Illinois Normal (SIN) in Carbondale	8
University of Illinois (UI) in Champaign-Urbana	12
John Christian Bailar, Jr.	14
Graduate Days at the University of Illinois	16
The Legacy of the Bailar Lecture	18
My Dissertation Research	21
Industrial Research—Rohm & Haas (R&H)	28

CHAPTER 2
MARY, THE CHILDREN, AND ME	33
Home in Evanston	35
Sabbatical Year in Denmark	37
Back Home in Evanston	60
Sabbatical in Rome	61

Contents

CHAPTER 3
FACULTY POSITION AT NORTHWESTERN UNIVERSITY (NU)   69
   September 1946   69
   Some Northwestern History   70
   Teaching   72
   Research   77
      Research Environment   77
      Research at NU   80
      Ligand Substitution Reactions of Octahedral Cobalt(III) Complexes (Co(III))   84
      Acid Hydrolysis or Aquation of Metal Ammine Complexes (Cobalt(III) or Co(III))   86
      Base Hydrolysis of Metal Ammine Complexes   88
      Linkage Isomers   91
      Ligand Substitution Reactions of Pt(II) Square Planar Complexes   94
      Organometallic Chemistry   100
      Synthetic Oxygen Carriers   109
   Chairman of the Chemistry Department   115

CHAPTER 4
OTHER ACTIVITIES   117
   National Academy of Sciences (NAS)   117
   American Chemical Society (ACS)   124
   Gordon Research Conferences (GRC)   136
   International Conference on Coordination Chemistry (ICCC)   144
   Funding Agencies   146
      The PRF Advisory Board   146
      North Atlantic Treaty Organization (NATO)   148

CHAPTER 5
COUNTRIES AND CHEMISTS VISITED   153
   Italy   153
   People's Republic of China   159
   Germany   168
   Australia   173
   Kuwait   177

CHAPTER 6
FOREIGN GUESTS HOSTED   183
   J. K. Beattie   183
   P. Dwyer   183

F. Lions	186
A. M. Sargeson	187
V. Gutman	187
S. Asperger	188
A. A. Vlček	189
C. E. Schäffer	190
C. K. Jorgensen	190
M. Becke-Goehring	191
E. O. Fischer	192
W. Klemm	192
V. Balzani	193
I. Bertini	194
F. Calderazzo	195
A. Ceccon	195
I. L. Fragala	195
R. Romeo	196
D. A. Buckingham	196
P. Sobota	197
B. J. Trzebiatowski	197
M. E. Vol'pin	198
L. G. Sillén	200
G. Schwarzenbach	201
L. Venanzi	202
K. B. Yatsimeski	202
C. Addison	203
J. Chatt	203
H. J. Emeleus	204
J. Lewis	205
R. S. Nyholm	206
A. J. Poë	207
M. D. Johnson	208

CHAPTER 7
EMERITUS PROFESSOR 209

APPENDIX 223
    Basolo Award Medal 224
    Honors 225
    Dedications, Memberships, and Publications 226
    Other Co-Authors 226
    Ph.D. Students 227
    Postdoctorates 277

Contents

Service to Chemistry	228
The Dream Team of NATO Workshop, 1972	230
Countries Visited	231
Academic Family Tree of Fred Basolo	232
NAME INDEX	233
SUBJECT INDEX	239

LIST OF ABBREVIATIONS

ACS	American Chemical Society
AEC	Atomic Energy Commission
AFOSR	Air Force Office of Scientific Research
BCST	Board on Chemical Sciences and Technology
BIP	Basolo-Ibers-Pearson
C&E News	Chemical and Engineering News
CCHS	Christopher Community High School
CCS	Committee on Chemical Sciences
CEO	Chief Executive Officer
CNR	Consiglio Nazionale delle Ricerche
DOE	Department of Energy
EDTA	ethylenediaminetetraacetate
ETH	Eidgenössische Technische Hochschule
EU	European Union
FRS	Fellow of the Royal Society
GNP	Gross National Product
GRC	Gordon Research Conference
ICCC	International Conference on Coordination Chemistry
ICI	Industrial Chemical Industries
INEOS	Institute of Organoelement Compounds
IRS	Internal Revenue Service
IUPAC	International Union of Pure and Applied Chemistry
JACS	Journal of the American Chemical Society

List of Abbreviations

JSPS	Japanese Society for the Promotion of Science
KTH	Royal Institute of Technology (Sweden)
MC	Mary Catherine (Basolo)
MIT	Massachusetts Institute of Technology
NAS	National Academy of Sciences
NATO	North Atlantic Treaty Organization
NIH	National Institutes of Health
NMR	Nuclear Magnetic Resonance
NRC	National Research Council
NSF	National Science Foundation
NU	Northwestern University
OSHA	Occupational Safety and Health Administration
PGA	Professional Golfer's Association
Ph.D.	Doctor of Philosophy
PRC	People's Republic of China
PRF	Petroleum Research Foundation
RC	Research Cooperation
R&H	Rohn & Haas Chemical Company
SIN	Southern Illinois Normal
S_N1	Substitution nucleophylic unimolecular
S_N1CB	Substitution nucleophylic unimolecular conjugate base
S_N2	Substitution nucleophylic bimolecular
SUNY	State University of New York
TUM	Technische Universität München
UCL	University College London
UI	University of Illinois
UK	United Kingdom
US	United States of America
USSR	Union of Soviet Socialist Republics
WPA	Works Progress Administration
WWII	World War II
WWW	World Wide Web

LIST OF FIGURES

1-1	My Parents	2
1-2	My Cousin Aldo and Me	4
1-3	My Mother	6
1-4	The "Four Horsemen"	9
1-5	Professors James Neckers and Kenneth van Lente	10
1-6	My Entire Family in 1939	11
1-7	Honorary Degree in 1984	12
1-8	Me as a Graduate Student	13
1-9	Professor Reynald C. Fuson	13
1-10	Professor John Christian Bailar, Jr.	14
1-11	Professor Alfred Werner	15
1-12	Professor Sophus Mads Jørgensen	18
1-13	Professors Alfred Werner and Arturo Miolati	19
1-14	R&H Baseball Team, 1943–46	31
2-1	Mary and I, First Christmas, 1947	34
2-2	Mary's Mother	35
2-3	Professor Jannik Bjerrum	37
2-4	MC and Freddie in Denmark	39
2-5	Children in Denmark	39
2-6	After Dinner at the Carlesberg Mansion	40
2-7	Jannik Bjerrum's Research Group in 1954–55	41
2-8	Book Cover of *Mechanisms of Inorganic Reactions*	42

List of Figures

2-9	Professor E. O. Fischer	44
2-10	Professor Walter Hieber	44
2-11	Professor Wilhelm Klemm	46
2-12	Professor Arthur Martell	47
2-13	French–Italian Border, 1955	48
2-14	Professor Lamberto Malatesta	51
2-15	Professor Luigi Sacconi	52
2-16	Professor Vincenzo Caglioti	53
2-17	The Royal and Ancient Golf Course in St. Andrews	55
2-18	Professor Clifford Addison	56
2-19	Professor Sir Ronald Nyholm and Family	58
2-20	Professor Joseph Chatt	59
2-21	My Family on Sabbatical in Rome (1961–62)	63
2-22	Book Cover of *Coordination Chemistry*	67
3-1	Professor Charles D. Hurd	70
3-2	Professors V. N. Ipatieff and Herman Pines	70
3-3	Walter Murphy	71
3-4	Professor Robert Parry and His Wife Marj	74
3-5	Professor Harry J. Emeleus	75
3-6	Professor Ralph G. Pearson and Myself	78
3-7	NU Inorganic Chemistry Faculty, 2001	79
3-8	Sir Christopher K. Ingold	80
3-9	Attendees at Symposium at NU, 1957	81
3-10	BIP Blackboard Demonstration Given by Bradley Holliday	82
3-11	Profesor Henry Taube and Myself	86
3-12	Professor Arthur Adamson	89
3-13	Professor John L. Burmeister	93
3-14	Professor Boguslova J. Trzebiatowska	95
3-15	Rates of Reaction of *trans*-PtCl$_2$(pyridine)$_2$	97
3-16	General S$_N$2 Mechanisms of Pt(II) Complexes	98
3-17	Professors Umberto Belluco and Marino Nicolini	99
3-18	Professor Luigi Venanzi	101
3-19	Professors Andrew Wojcicki, Harry Gray and Myself	103
3-20	Professor Fausto Calderazzo	108
3-21	Professor William C. Trogler	108
3-22	Professor Qi-Zhen Shi and Me	109
3-23	Professor Brian Hoffman	112
3-24	Pearson's and My Research Groups	115
4-1	Professor Kazuo Saito	118
4-2	Professor Shoichiro Yamada	119

List of Figures

4-3	Professor George Pimentel	122
4-4	*Chemical Abstracts Index*	124
4-5	Mary and Me in 1983	129
4-6	Arnold Thackray	131
4-7	*C&E News*, ACS President	132
4-8	Directors of GRC	137
4-9	Professor Kim Dunbar	140
4-10	Dr. Nadine de Vries	140
4-11	Conseil de Physique Solvay Conference, 1911	142
4-12	Professors Stanley Kirschner and Jan Reedijk	145
4-13	Professor Melvin S. Newman	147
4-14	Professor Renato Ugo	150
4-15	Professor Robert Burwell	151
5-1	Concorsa, Rome, 1962	155
5-2	Professor Giulio Natta	156
5-3	Foreign Membership in Italian Academy of Science Lincei	157
5-4	Professor Luigi Sacconi, Lamberto Malatesta, and Me	157
5-5	*Laurea Honoris Causa*	158
5-6	The Sgarbi Family, My Relatives in Italy	159
5-7	Professor Yun-Ti Chen and Me	160
5-8	200 Faculty at Nankai University, 1979	161
5-9	Class on Coordination Chemistry, Nankai University, 1979	162
5-10	Vice Premier Fang-Yi, Mary, and Me	165
5-11	Professor Egon Wiberg	169
5-12	Professor E. O. Fischer and Me	170
5-13	Professors F. Albert Cotton, Andy Wojcicki, and Geoffrey Wilkinson	171
5-14	Professors E. O. Fischer and Geoffrey Wilkinson	171
5-15	Kuwaiti Students	178
5-16	Me in Kuwaiti Robe and Head Scarf	180
6-1	Professor Frankie Dwyer	184
6-2	Character Sketch of Frankie Dwyer	186
6-3	Professor Francis Lions	187
6-4	Professor Viktor Gutman	188
6-5	Professor S. Asperger	189
6-6	Professor Antonin Vlček and his Wife	189
6-7	Professor Margot Becke-Goehring	191
6-8	Professor Mark Vol'pin and Me	199
6-9	Professor Lars Gunnar Sillén	200
6-10	Professor Gerhart Schwarzenbach	201

List of Figures

6-11	Professor Yatsimerski	202
6-12	Professor Anthony Poë	207
7-1	In Front of My Brother Martin's Grocery Store	210
7-2	Sgarbi Brothers in Milan	213
7-4	Mary and Me in Copenhagen, 1995	217
7-5	Willard Gibbs Medal, 1996	218
7-6	Priestley Award, 2001	219
7-7	On My Scooter in the Tech Building	220

1

FROM COELLO TO INORGANIC COORDINATION CHEMISTRY

MY PARENTS

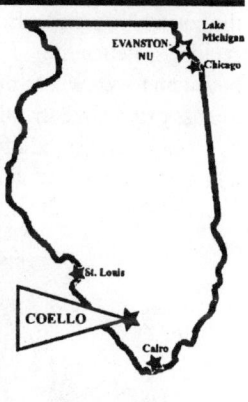

During these many years when I have mentioned Coello* to my friends, the questions asked of me have generally been: What is Coello? or Where is Coello? Let me begin by telling you that Coello is a small (average population 500) mining village, a suburb of Christopher (pop. 3,000), about 25 miles north of Carbondale, Illinois—home of Southern Illinois University (see map). I start with this account of Coello because, as will be seen, it has had much to do with my approach to living throughout my life. It was there that I learned how to respect and understand others, no matter their status in life.

Before describing my childhood, I must say something about my parents. They, like many other Europeans, emigrated to the United States in the early 1900s. They, like

---

*Illinois State and Franklin County had named and registered the name of North City for the small mining village. The name Coello resulted from the people unanimously wanting to honor their first Postmaster, whom they all liked, by naming the town after him (Peter Coello). They told him that they would do this, providing that he pay the $25 fee to register the name at the Franklin County City Hall. He did this, and everyone in the area knew the village as Coello. Signs on the highways, however, indicate North City, not Coello. Mail is correctly delivered with either of the two names being used. Because of the rowdy behavior at the weekly dances held in the town, it was also dubbed "Pistol City." It is, perhaps, the only town that has three names, all of which are recognized by people in the area.

the early 1900s. They, like others, came because of the near-starvation conditions existing in Europe at that time. My parents' circumstances were similar to the other immigrants of the time. Therefore, there is really no reason to tell the story of my parents, for it generally is the same story told by others of that status abroad who were fortunate enough to reach the United States. However, my parents, more than anything else, helped to shape me into the person I am today and I feel certain that this has nothing to do with genetics, but all to do with their setting an example for me to follow. Therefore, I feel it only right that they be given some time and space in this, my autobiography. They certainly are my two role models for life in general (**Fig. 1-1**). Later, my model as a chemist will be mentioned.

My mother (Catherina Marino-Basolo) came from the Piedmont region of northern Italy. Her family had eight children, six of whom survived. They lived in a small peasant country area called Tetti, near the village of Dronero. Mother told me many things about her family life in Tetti. The following most readily comes to mind. She stated that they survived by eating mostly "polenta" (a sort of hard corn meal mush). Meat was served only once or twice a year, whenever a chicken happened to die. In winter, their house was largely heated by the body heat of their cows, which slept in a room attached to the house. No wonder Europeans living under these conditions came to the United States seeking a better life! I have been fortunate in having had the opportunity to visit the house and area where my mother was from. Even after seeing it as it is now, it is only too easy to realize that the accounts my mother gave of her life there were true.

My father's (Giovani Basolo) situation in Italy was much better. He came from a family of three boys and one girl. His father was a cow merchant who was considered a moderately wealthy man in his village of Borgiallo. Being the oldest child, my father

Figure 1-1. This is the picture that I like best of my parents, for I recall their life style after the retirement of my father from his work in the coal mine. They are my first role models.

was given the little money they had to come to the United States, traveling in steerage class, as did most emigrants. This was common practice of the time: whatever monies were available within a family were given to the oldest son to travel to America. He would then work in the United States and save his money until there was enough to send for his next oldest brother. My father was not married when he left Italy. I am grateful that I also had the opportunity to visit his small village which is located atop a hill, thereby providing a good view of the surroundings below. It did appear that my father had slightly better living conditions than did my mother.

Upon arriving in America, my father traveled to Coello, where most of the men were immigrant coal miners. He and the other miners lived in the company houses that were made available by the mine owners. Two workers shared one room. Fortunately, there was enough money in my father's family that he did not have to return any of his earnings to fetch his brother. My father's roommate (since then my uncle) asked him if he would be willing to help bring his sister to the United States. This request was made because my uncle could not earn enough to bring both his next oldest brother and his sister. The proposition made to my father was that if he paid for the sister's travel, and if they liked one another enough to be married, my uncle would not owe my father anything. If they did not like each other, my uncle would, in time, pay my father the travel cost of his sister. All went well, my uncle had nothing to pay, and my parents lived a happy life together with three of us children. They both died of cancer after a long painful illness. My mother died at age 59 and my father at age 62. Keep in mind that life expectancy, at that time, was much lower than today.

THE EARLY YEARS

I will always remember my years as a child and youngster in Coello. These years were, in my opinion, the foundation that made me an honest, hard working, dependable, and credible person. Regardless of early hardships, I have always considered my early childhood to have been the most enjoyable and most instructive period of my life (**Fig. 1-2**).

I was born on 11 February 1920, just before the devastating depression of 1930. My sister was nine years older than I and my brother was eleven years older. At times during dinner, when different views on a particular subject were discussed, I would often speak up to offer my opinion. My brother or sister would frequently interrupt me saying, "You keep out of this, because you were a mistake!" My mother would blush, saying, "Don't say that—it is not true." However, this being before the days of contraceptives, there must have been many unplanned births. Nevertheless, I do believe that I was a mistake, and should not be here. In spite of the various effects that psychiatrists imply this may cause, I believe that it has had no such effects on me.

When I was six months of age, my parents decided to return to Italy, where they planned to help my grandfather sell cows. This would be a much less stressful and dangerous job than working in a coal mine. Although Catholics are baptized soon after birth, knowing that they were to return to Italy, mother wanted me to be baptized in the

Figure 1-2. I am to the left with my cousin Aldo. Note our bare feet in the summer to save wearing out our shoes.

same church where she had been baptized. I have since been back to Italy many times, and have seen this old church. I was told that mother spent most of the six months in Italy crying that she wanted to return to the States. My father then returned to his job in Coello, and I celebrated my first birthday aboard ship, as we returned.

Most of the coal miners in Coello were immigrants from Italy, and most of them were from the Piemonte region and spoke the Italian dialect "piemontese." One could go for days in the neighborhood speaking only "piemontese." Most of the immigrants had no formal schooling beyond the third or fourth grade, but this is not to say that they were not intelligent.

The first thing my parents did was to buy a plot of land of about 4-5 acres. The second was to plant grapes on their land. Then they waited for a few years until the grape vines began to bear grapes and when they had saved enough money, they built their house. This was done during the summer, when the mine was closed. All of these coal miners were also common laborers, capable of doing the work required to build a house, but none were architects. Thus, the houses were built as square boxes, divided into four rooms. The houses had no central heating nor plumbing and, of course, no air conditioning. The use of an outdoor toilet in the winter and the taking of a bath in a large aluminum tub were challenges. The house was heated in only one room by a coal stove, which was also used for cooking. At times when my mother was cooking, she may have had more than one item that demanded her attention. As a child, I would often be asked to stir some item as it cooked so it would not stick to the bottom of the pan. These experiences gave me confidence in the kitchen, and has made cooking one of my hobbies, resulting in my being able to cook some very good Italian dishes. I do this without recipes, using only my recollections of what my mother did.

The Italian emigrant coal miners in Coello never went on vacation, but spent the summers catching up on chores which needed to be done at home. They all found it necessary to plant a sizeable garden—this was not a hobby but was needed for food throughout the year. I recall vividly our harvesting tomatoes, beans, cucumbers, etc., then sitting under the shade of a tree to can them. I have already mentioned how the miners planted grape vines and later built their houses. This was all done by a group of friends and neighbors helping one another. In September, when the grapes were ready to pick, they would make the rounds, picking and smashing the grapes with an appropriate grinding wheel at the various houses. I think the smashing of grapes with bare feet to make wine must be a myth propagated by movies, using a beautiful girl to do the smashing with her bare feet. Actually, the mashed grapes are put in a large cistern and allowed to ferment for one or two weeks. During this time, the grapes are submerged under the grape juice once or twice. Finally, the wine is drawn off the bottom of the cistern and put in 50 gallon barrels to age for a few months.

As a child, I always looked forward to these gatherings of the miners at our house. Once the grapes were picked and crushed, the miners would eat and partake of wine made from the previous year. This would be followed by one or two forms of entertainment. One was the singing of Italian songs, always sung loudly and clearly. Their voices were far from the best, but they usually started with singing. The remainder of the evening was spent playing cards, a hand game, or "boccia." The card game "Tarocchi" is a game with cards from Italy that are not available in the US. The hand game may also be played only by Italians. It involves two opponents putting up fingers of one hand while they simultaneously shout a number. If one of the participants shouts the correct total, he wins a point. This game can get very noisy because some players are able, at times, to hesitate slightly, getting a glimpse of his opponent's fingers in time to adjust the number of his fingers and shout the correct total. This, then, often resulted in a loud argument in Italian which, as a child of about 10, I enjoyed hearing. These exchanges were responsible for my early learning of many obscene Italian words. The other game played was boccia, which is reasonably known in the US. It is a game that resembles the much more elegant British game of bole. The miners would often end up drinking wine somewhat late into the night. I would stay up in order to hear the many stories about their lives in Italy and Coello, until mother had me go to bed. Now, having all of this to recall is great fun for me.

Most families also had one or two cows, a family pig, and chickens. The friendly pig had to be killed each fall and was replaced by a young pig to repeat the cycle. This was done to provide meat in the form of sausage, and other items, such as all the organs, blood, head, ears, and feet. It was said that the only thing that was not eaten was the pig's squeal! Eating all of these items as a child is the reason I now find dishes, such as brains and other organs, which most people will not even taste, to be delicious.

COELLO ELEMENTARY SCHOOL

The great depression of 1930 arrived when I was ten years old. This meant that I was too young to go to work, so my elementary schooling continued uninterrupted. Not so with my brother and sister, who were a decade older. They had to go to work in low paying jobs to be of help to the family. This meant that they did not even get to go to

Figure 1-3. My mother, Catherina Basolo, who was also my best friend. I loved her dearly.

high school, as was likewise true of many other youngsters about their age. However, my first job was at age 15, when I picked peaches from sunrise to sunset for the rate of 25 cents an hour. The most annoying thing about this was trying to wash away the sticky, irritating peach fuzz which got all over one's body.

The starting age for elementary school was six, but my mother (**Fig. 1-3**), believing that I was ready for school, managed to have me start at age four and a half. Since we always spoke "Italian" at home and in the neighborhood, I understood but spoke little English. Mother told me that if I was asked my age, I was to hold up one hand and a finger. It worked so my mother must have been correct, because I had no problems keeping up with the class work. One other thing my mother did forever changed my name from what is on my birth certificate. It is recorded as Alfredo (Alfred), but I was having a problem spelling it, so mom said, "Just write Fred." Clearly, I have never changed it.

However, there was one problem she did not consider, which was that, for children in their teens, 1 to 2 years of age makes a huge difference in their behavior. I always enjoyed playing outdoors, and during the summers I played baseball almost every day. Tennis, golf, basketball, and football were all too expensive for Coello. I was able to compete at varsity baseball, but the only varsity sports at my high school, Christopher Community High School, were football and basketball. I was too small to even try out for either of these varsity teams. One other problem involved girls. Being a year or two younger makes it quite difficult for a boy to have a girlfriend who is in his same school class. For example, I did not attend the junior-senior prom as a junior. I did, however, attend the prom my senior year because I was asked to go by a very delightful girl who did not have a boyfriend, and would have otherwise missed the prom.

The elementary school which I attended was a four room building in Coello. Each room was divided into two classes. With only one teacher per room, this meant that a class would get only half the schooling that it would with one teacher per class. I do not recall very much about my years in elementary school. One teacher whom I recall best was my first and second grade teacher, Mrs. Shanon. She was very kind and seemed to love each and everyone of us in her class. Just the opposite of her was Mr. Arterberry, my seventh and eighth grade teacher. He was very demanding, and succeeded in frightening many of us. What I do recall most about elementary school is running most of the way home to listen to "Jack Armstrong, The All American Boy." This was a radio program and it always ended with Jack in some deadly serious situation. I had to run the mile home along the railroad tracks to get there in time to hear how he managed to once again escape.

CHRISTOPHER COMMUNITY HIGH SCHOOL (CCHS)

As mentioned earlier, my brother and sister, and most other children of their age in Coello, had to go to work during the depression and did not get to attend high school. However, about ten years later most of the children in my graduating class did attend CCHS. This was about two miles from Coello, and there were no school buses. We had to walk in all kinds of weather, almost one mile through fields, for using the one available road would have added another mile to the distance.

The high school, with its ca. 300 students, 15 teachers, library, football field, basketball court, etc., was really "big time" to those of us from Coello. In spite of this, we did adjust and most of us enjoyed our four years there. In fact, what remained of our class of 1936 celebrated its 50th anniversary, and attendance was excellent. However, we hardly recognized one another in our old decrepit states.

The high school was sufficiently small that the faculty had an opportunity to get to know most of the students. The principal was Mr. Hughes, who had previously been the superintendent of all elementary schools in Franklin County. When he visited the elementary schools in the county, he tried to get to know some of the students. Fortunately I was one of the students he took an interest in, and he remembered me and talked with me at times in high school. Hughes was very interested in having his better high school students continue their education, although, at that time, many of the graduates from CCHS did not go to college.

My first contact with chemistry was in high school. At the time, I really did not even know the meaning of the word. We children certainly did not grow up with a chemistry set—with our poverty, we even saved our shoes for winter by walking barefooted in the summer. In my senior year of high school, it was my good fortune to have a beautiful young blond lady as my chemistry teacher. This was her first year teaching, having just graduated from college. She was also the only person on the teaching staff who had at least one year of chemistry, which was required for her degree in home economics. At our first class meeting, she told us that she hated chemistry and did not know much about it, so we were going to be largely on our own. What we would learn was to come mostly from our textbook and our laboratory manual. She said that she would open the laboratory on Saturday mornings and be there to help take care of anyone who got injured. This approach to teaching chemistry gave me the freedom to concentrate on what interested me in the textbook and, even more so, to get very excited about the laboratory experiments. Needless to say, this was the course I liked most and remember best, and the course that made me feel I would want to take more chemistry courses. That my teacher was a beautiful young blond may also have contributed somewhat to my interest in chemistry. Enough said for those of us chemists who state that few students select science, particularly chemistry, as a career because of their high school teachers. We contend that high school teachers of science and mathematics do not know enough about the subject to be teaching it, and do not like it enough to motivate students in that direction. Just the opposite was true in my case.

Some 25 years later I was teaching a large (approximately 500 students) freshman chemistry class at Northwestern University (NU). I always encouraged students who had questions on the lecture to stop by afterwards and ask me or come to my office. After two or three weeks of lectures, a student came up to the front of the lecture room on his way out to ask me if I had attended CCHS, and if I had had a teacher named Vivian Wells. I answered, "Yes, but how do you know all of this?" He replied, "She is my mother, and wants me to give you her regards." I invited him to bring his mother to my office the next time she was on campus visiting him. Later, he told me they had come by to see me two different times, but I was not in my office. That was when I was doing a lot of traveling as the President of the American Chemical Society (ACS) and giving many invited lectures on our research. Although I missed seeing her, in retrospect, I think it better to just remember her as my beautiful young first chemistry teacher.

SOUTHERN ILLINOIS NORMAL (SIN) IN CARBONDALE

My parents wanted very much to have me continue on to college, but they were financially unable to help me. Fortunately, President F. D. Roosevelt had started welfare programs, such as the W.P.A. One of the programs was for college students to receive $25 per month as payment for some work on campus. This was not much money, but I was able to enroll in Southern Illinois Normal (SIN), which is only 30 miles south of Coello, and was one of the least expensive colleges. My first duties were to clean all the classroom blackboards and make certain there were chalk and erasers in the large main building in time for 8:00 AM classes. In my sophomore year, I was promoted to working in the library. As a chemistry major, during my last two years, I worked in the campus doctor's office, doing blood and urine analyses. I was also able to supplement my regular income by living free in an attic apartment of the Dean of Women's house in exchange for taking care of the yard in summer and the coal furnace in winter. I also had one other job on Saturdays, working in a men's store. Every other week I would hitch-hike a ride home to bring mother a bag full of dirty clothes, and return with clean clothes and some food that mother knew I liked.

I recollect two interesting things that happened when catching a ride in the dark on Saturdays after work. In those days of few cars, it was easy to catch a ride, particularly if it was obvious that one was a college student. One evening a car stopped to pick me up, but the driver asked me to use the back door behind the driver. It was clear why he asked me to do this when I got in the car. The front seat had been removed in order to place a long flat object there. The object was covered with a blanket, and stretched from the front seat to the back. After we had gone a few miles, sensing I was curious as to what the object next to me might be, the driver reached over to pat it and said, "Don't worry, he is good and dead. He will not hurt you." He told me that he had gone to a mental hospital in Cairo to get the body with his car because their ambulance was being used for another purpose.

The other incident illustrates that I was the only person from Coello going to college (now I am the only person from there with a Ph.D.). Again, I managed to catch

From Coello to Inorganic Coordination Chemistry

Figure 1-4. "The Four Horsemen" is the name we chemistry majors at SIN initiated for our four faculty in 1939.

a ride, put my luggage in the back of the car, and, this time, sat up front next to the driver. After dark, he stopped and dropped me off in the middle of nowhere. I had to walk about two miles through some fields to get home. However, the driver later stopped at a gas station, and discovered that I had taken his luggage by mistake. He then asked the attendant at the gas station if he knew of a SIN student who lived in Coello. The attendant answered, "Oh, yes, there is only one, the Basolo kid." He gave good directions to the driver, because the driver soon arrived at our house to exchange bags. Other than being angry for all the trouble that this had caused him, he said, "No problem. Just hang in there and some day others from here will follow you to college." This illustrates the friendly understanding of people in that general area of Illinois.

As a Normal school, SIN was intended to prepare students for a teaching career. Those students who attended for two satisfactory years would receive a certificate qualifying them to teach in an elementary school in the State of Illinois. Those who successfully finished a four year program received a B.Ed. (Bachelor of Education) degree. This qualified them to teach at the high school level in Illinois. In the general area where we lived, a high school teaching position was considered to be a respectable, moderately high paying position. Such a position near the area where we lived was what my parents and I looked forward to, but clearly this did not happen.

My freshman year at SIN, I took chemistry, mathematics, English, and some education courses. I continued to be excited about chemistry, so I finally obtained my B.Ed. with majors in it and in math. I had a year of physics, but did not like the instructor nor the course, so the one year was all I took. Just the opposite was true in chemistry. My freshman general chemistry course was taught by Professor James Neckers, an outstanding teacher, whose interest in students was paramount. His Ph.D.

was in analytical-inorganic chemistry. The lab work he had us do was mostly qualitative and quantitative inorganic chemistry, both of which I found very challenging. Jim was an outstanding lecturer, and with our small classes, we were able to ask questions and get involved in short discussions. He was also the chairman of the four man chemistry department. Our chemistry majors group, called *Chemeka*, decided to refer to them as "The Four Horsemen" of Notre Dame fame (**Fig. 1-4**), at a time when Ronald Reagan was not yet president, but only a football player. This name has caught on, and continues to be used.

My other courses and instructors in chemistry, in chronological order, during my remaining three years, were Professors Talbert Abbott in organic chemistry, Kenneth van Lente in physical chemistry, and Robert Scott in biochemistry. Abbott gave good lectures which followed, almost verbatim, the outstanding textbook written by Professor J. B. Conant, who later became president of Harvard University. One of the students in our class always read ahead in the book to know what would be presented at the next lecture. Thus, he came prepared to ask questions and dominated any discussion during the lecture, so we called him "J. B." Ken van Lente was tough and very demanding in his course in physical chemistry. Physical chemistry was my most difficult course, but I did come to like Ken's approach to the course. Professor Scott, in biochemistry, was outgoing and less interested than the other three faculty in his course and in his students. He was much more interested in his quail hunting dogs, which were among the best in the US in hunting competitions. However, Scott did do something in his lab to get us started on a urine analysis that got our attention and that we students will always remember. He had a small sample of urine and said that before doing the analysis, we should always make certain that it is urine. This, he said, should be done by putting one's finger in the urine and tasting it. He proceeded to demonstrate, using the urine sample, in which he placed one of his fingers and then put another finger in his mouth. He asked us students to do the same, in order to know if we should proceed with the analysis. Almost all of us did it, and we can say that urine does not taste too bad. He then informed us that he had used a different finger to put in his mouth than the finger that he had put in the urine, and he told us how important it is to carefully observe the experimental results in the lab.

Figure 1-5. Professors James Neckers (at left) and Kenneth van Lente (1903-98), both at SIN; my very best teachers and friends at SIN.

Abbott and Scott died early, but van Lente and Neckers lived in good health to a fine old age. Ken died this past year at age 94, and Jim still lives at age 98 (**Fig. 1-5**). I

Figure 1-6. My entire family in 1939.

had lunch with Jim on Memorial Day weekend in 2000. For years, I have taken Ken and Jim to lunch when I go to Coello ("God's Country") to visit my parents' graves. I started referring to that area of Southern Illinois as God's Country to my children but, as they grew older, they said I called it that because only God could love it.

My final comment on my undergraduate days at SIN has to do with a conversation I had with Neckers during my junior year, which changed my entire career expectation of becoming a high school teacher. Jim told me I was doing well scholastically in chemistry, and said I should seriously consider going on to graduate school. My first reaction was, "What is graduate school?" Then, there were my parents' opinions to be dealt with (**Fig. 1-6**), as they were expecting me to finish school after four years at SIN and return home to a position of high school teacher. I would have been happy doing this, for I did enjoy teaching. I know this because, in one of the education courses at SIN, I was required to practice teach in a high school course in my chosen major. My practice teaching was done in chemistry under the supervision of an old high school teacher. In his twenty or more years of teaching, he had not kept up with developments in chemistry, and had forgotten much of the older material that he perhaps knew at one time. In any case, I was sure that I knew more chemistry than he did, yet at times it was more prudent to do it his way. I did survive, and when I told Neckers about my experience with the high school teacher, he said this was generally known, but since he had only a few more years to retirement, they had decided to let well enough alone. One other point that can be made about my B.Ed. which qualified me to teach high school chemistry in the State of Illinois is that it strengthened my feeling that I would want a teaching position.

SIN was responsible for me getting an education beyond high school, because it would have been impossible for me to go elsewhere to college. The advise I received from Professor Neckers is the reason I went on to get a Ph.D. in chemistry. SIN, now SIU, has been good to me over the years, honoring me with all their prestigious awards to its alumni, the most prestigious award being an honorary degree in 1983, when I had to give the Commencement address (**Fig. 1-7**).

My love for teaching has been reflected by receiving high evaluations from students in my courses at NU and receiving two of the highest Awards for teaching and chemical education given by the American Chemical Society (ACS). This was accomplished while also receiving the highest ACS Awards for research in inorganic chemistry. I am honored and pleased to have received the highest ACS Awards for both teaching and research (see Vitae in Appendix). This is seldom done by faculty in research-oriented universities, such as NU, where there is little time for teaching and where faculty promotions depend mostly on their research and its funding.

UNIVERSITY OF ILLINOIS (UI) IN CHAMPAIGN-URBANA

I did take the advice of Neckers to continue on to graduate school, and the rest of my career is history. Fortunately, I was offered a teaching assistantship at the UI. The stipend offered to help me continue my education was $60 per month (much more than the $25 received at SIN), which seemed like a lot of money at the time. Needless to say, I accepted the offer in 1940 and obtained my Ph.D. in 1943 (**Fig. 1-8**). However, before going to the UI, I decided to learn as much about it as possible. What I learned

Figure 1-7. SIU awarded me an Honorary Degree in 1984, and had me give the commencement address.

was that its faculty included the giants of organic chemistry, such as Professors Adams, Fuson, Marvel, and Shriner. Inorganic chemistry consisted of Professors Audrieth, Bailar, Moeller, and some younger faculty. The late Professor B S Hopkins, known for his research effort on the separation of rare earths by fractional crystallization, had just retired. Unfortunately, he was of the honest opinion that he had discovered element 61 and called it Illinium. During the years of my graduate studies at the UI, all of the large periodic tables on the walls of the lecture rooms had the symbol Il for element 61. This was later found to be incorrect, and the element finally found in the fission products of U^{235} was given the name promethium and the symbol Pm. The periodic tables at the UI have long since been changed to Pm for element 61.

Figure 1-8. Me as a graduate student (1940-43).

When I arrived at the UI, I was overwhelmed with what to me seemed to be the large size of the twin cities and of the UI campus. On arrival, I had to make a choice of what subdiscipline and faculty mentor I would want, and this was frightening. It was easy to rule out physical chemistry and biochemistry because I had not liked either at SIN. I did not dislike organic chemistry and I decided to talk to two of the organic professors about their research. One seemed to want to tell me about his worldwide fame, rather than discuss his research and the nature of the problem I might work on if I joined his group. It was easy for me to rule him out as my possible mentor. The other organic chemist I talked with was entirely different. He told me about his research work, as well as what might be my dissertation problem. This was Professor Reynald C. Fuson, an outstanding chemist, and a very fine person (**Fig. 1-9**). I would have been delighted to have had him as my mentor. However, because of the excellent teaching of Neckers at SIN, I was of the opinion that inorganic chemistry was to be my area of chemistry, providing I found a suitable dissertation advisor.

Figure 1-9. Professor Reynald C. Fuson (1895-1979) at the UI. An outstanding organic chemist, an excellent teacher, and a very good friend.

The first professor of inorganic chemistry whom I talked with left me with the feeling that he would be a difficult person to work with. The second, Professor Therald Moeller, told me about his research interests in the chemistry of the rare earths, which did not excite me. However, he was a quite easy going person, whom I could have worked with as my dissertation director.

Professor John C. Bailar Jr. (**Fig. 1-10**) was the last faculty member I talked with. He impressed me with the knowledge of his research in progress, and he was very excited about the research he suggested I might do for my Ph.D. dissertation. Since most of the work he and his students had done dealt with cobalt(III) complexes, he suggested to me that we should have a look at platinum(II & IV) in the same manner as his work on the cobalt(III) systems. I was intrigued by the dissertation research he suggested. Here was a professor knowledgeable in organic chemistry doing work on inorganic which, to me, seemed like a win-win situation. In addition, he seemed to be a caring person, with a good understanding of people and their needs. My choice of John as my dissertation mentor was one of the very best decisions I have ever made. He was, in the true sense, my "Professional Father." Not only was he helpful to me during my dissertation research, but I went to him for his advice on professional and personal matters until almost the time of his death in 1991. Because of his very good chemical research, his outstanding teaching, his tremendous service to chemistry, and his personal qualities as a caring, honest, understanding human being, he is my second role model after my parents. He and his wife treated all of his graduate students as family. Since Bailar became a friend and I owe so much to him, I want to say a bit more about him.

Figure 1-10. Professor John Christian Bailar, Jr. (1904-91) at the UI. He was a very good inorganic chemist, an outstanding teacher, and excelled in all the personal qualities desired of a person. This is how he looked when I was a graduate student. He is my second role model.

John Christian Bailar, Jr.

John Christian Bailar, Jr. was born on May 27, 1904, in Golden, Colorado. He did his undergraduate work at the University of Colorado, Boulder. His father taught chemistry at the Colorado School of Mines, so that John, Jr. had an early exposure to chemistry. He attended the University of Michigan, where he obtained a Ph.D. in 1928, working with the organic chemist Moses Gomberg (1866-1947) as his mentor. John then joined the faculty at the UI as an Instructor at a salary of $2,100 per year. His initial assignment was to teach freshman chemistry which, at that time, was composed almost entirely of descriptive inorganic chemistry. Since he was teaching inorganic chemistry, he felt it necessary to do research on inorganic problems although his Ph.D. was in organic chemistry. Others have made such a change, even Alfred Werner. Now that inorganic chemistry has become extremely important, we see many of our younger chemists doing research in the areas of organometallic chemistry, bioinorganic chemistry, and new materials, all of which make use of the coordination theory.

Bailar prepared some organometallic compounds in an attempt to separate the rare earth elements, in collaboration with Professor Hopkins. Although this project was not successful, John believed that he should do research on an inorganic problem. After some time in the library, he found that papers by Alfred Werner (**Fig. 1-11**) were of interest to him. These Werner metal complexes, with geometric and stereo isomers, could be studied in a manner similar to that of organic compounds. In fact, John did such a good job with his research on these systems that he is viewed by chemists as the "Father of Coordination Chemistry in the US."

Figure 1-11. Professor Alfred Werner (1866-1919; Nobel Prize, 1913) at the ETH. After 100 years of not understanding metal complexes, he was the genius who gave us his coordination theory. This explained the bonding and structures of these compounds. His fundamental theory is now known to be the basis of most of the chemistry of metals.

Bailar was known and respected worldwide for his research, teaching, and service to chemistry, and to science generally. His awards are far too many to list here, and several are from foreign countries. Perhaps the one he may have cherished most was that given to him for his inestimable contributions to coordination chemistry by the Schweizerische Chemische Gesellschaft in Zürich. This award ceremony was at the celebration of Werner's birthday in 1966, when John was presented the only Werner Gold Medal ever to be awarded. He received the highest Awards of the ACS for his research and for his teaching. Bailar served on many important national and international committees, and he was very active in the ACS, serving as president in 1959. The ACS is the largest scientific society in the world, now with a membership of ca. 165,000. Perhaps the largest disappointment to Bailar was that he was not elected to the National Academy of Sciences (NAS). Those of us who knew John's research, and who now are members of the NAS, know that he deserved to be a member. Unfortunately, at the time, so very few of the members were inorganic chemists that it was almost impossible for one to get the votes required for membership.

GRADUATE DAYS AT THE UNIVERSITY OF ILLINOIS

Now, back to my years (1940-43) of graduate studies at the UI. One of the requirements to enter the Ph.D. program in chemistry was to pass a comprehensive exam in each of the four areas of chemistry (analytical, inorganic, organic and physical). I was certain that I would fail, because I had only done good to average work in my courses during the year and because many of my fellow graduate students arrived better prepared than I. In those days, organic chemists had yet to do very much on the mechanisms of reactions and syntheses of organic compounds. Organic chemistry was taught in terms of name reactions such as Diels-Alder, Grignard, etc. It seemed to me there were hundreds of these names that we were going to be responsible for on our qualifying exam. I decided that the only way I could learn enough of these to pass was to prepare index cards with the name on one side and its reaction on the other. Therefore, at the start, I had a large deck of cards to go through in order to memorize names to reactions and reactions to names. I would put the ones I knew in one stack and the ones I did not yet know in another. I had such a large number of index cards that this was taking a long time. I finally decided that once the two stacks were of the same size, I would stop with a 50/50 chance to pass. Later, I was surprised and very pleased to learn that I had passed the organic exam as well as the exams in each of the other three subdisciplines.

There was one more hurdle to pass before being admitted to the Ph.D. program, and that was an exam in one foreign language, which was usually German or French. My choice was German, because I had two years of it as an undergraduate. The article I had to translate from German to English had to do with the manufacture of sulfuric acid and some of its commercial uses. I translated the paper as best I could, but I knew enough about the chemistry involved that I could see that my translation was very poor, and I was certain that I had failed. Since we had to pass a foreign language before being allowed to start the second year, I felt sure I would not be allowed to continue. There was one week before the French exam was to be given, so I decided to register for it. I had never had any French, but I thought that my knowledge of Italian would help me more in French than German. I also found a book on chemical French which was very helpful. I worked with this book almost day and night for the one week before the exam. I took the exam and felt I had done a better job with the French than the German, because the chemistry of the translation made more sense than did that for the German translation. A few weeks later the list of students with passed or failed was posted, and, much to my delight, I had passed both exams. Later I learned that almost all students pass these exams, because there was already a tendency for all the best scientific papers to be published in English. Now language exams are no longer required of graduate students in chemistry, because all international chemistry conferences are presented in English and most of the chemistry journals are written in English.

My first year in graduate school I recall taking courses in coordination chemistry, organic chemistry, and ceramics. At that time, one was required to take a course from some science department outside chemistry, which is the reason I took ceramics. It turned out that I did not like the course, and I can recall very little about it. Qualitative organic chemistry was first rate. What organic chemistry I know, I learned in this course.

It was largely a laboratory class, where students were challenged to determine what compound, unknown to them, was in the sample. Many of us had to work for hours to identify our unknown sample, but we enjoyed the challenge and the final success was rewarding. Lectures in the course by Fuson were outstanding. He was a true professional and gave very formal lectures, following a well organized presentation of the important points he wanted to make. One thing that startled all of us 70 or so students in his class was that after one week of class he knew each of us by name. In my case, since I have an Italian name, he approached me by speaking Italian, and he was surprised that I understood. He then had me speak Italian and learned that I spoke the peasant dialect of northern Italy, which I had learned, as a child, from my parents. He said, "What a shame—you must learn the real Tuscan Italian. I will invite you to my apartment, where I will teach you while we have a good glass of Italian wine and listen to excellent Italian music." We did this, which was most enjoyable, but my Italian pronunciation did not greatly improve. For years, when we sometimes met at ACS scientific meetings, we would reminisce about my days at the UI.

John Bailar also was such a good person that he just had to be one of the very best teachers on the UI faculty. This was true for his teaching at all levels, from general chemistry through graduate courses. For example, he did many lecture demonstrations, which his beginning students enjoyed and which motivated them toward chemistry as a result. One of the experiments involved the violent reaction between a mixture of hydrogen and oxygen when initiated by an electric spark. Bailar demonstrated this by using a small homemade "cannon" containing a mixture of the two gases, an electrical device to cause a spark, and a stopper at one end of the cannon. The spark set off the explosive reaction and the rubber stopper was projected up into the vacant center of the lecture room. On one occasion, the device was mistakenly aimed incorrectly and hit a beautiful young lady in the front row. Bailar was so concerned that he was most caring and helpful to the student, who later became his charming wife Florence.

Bailar's graduate course on coordination chemistry was so well regarded by graduate students that, although there were only about six students in inorganic chemistry, the class usually had about thirty students enrolled. Because of all the library work John did, he also became interested in the history of this area of chemistry, starting with the serendipitous discovery of $[Co(NH_3)_6]^{3+}$ by Tassaert in 1797. Bailar then led his class through some of the more important experiments done over a period of 100 years, during which time many such compounds were prepared. However, the theories of the nature of these compounds were always proven wrong and these compounds were called "metal complexes," a term still used generically. It should be noted here, as I have always told my students, a scientific theory can be proven wrong, but it cannot be proven correct. This is because some new experiments may be done which prove the theory wrong, or some other theory may be proposed that also is consistent with all the known facts. Finally, after a century of curiosity about the nature of these compounds, it was left to the genius of Alfred Werner in 1893 to propose the *coordination theory*. Even with our present sophisticated instruments, the theory has withstood the tests of time. Since it is not possible to prove a scientific theory is correct, the Werner theory may not be correct, but it works and we believe it and we use it. Much the same can be said of the atomic theory of matter.

The Legacy of the Bailar Lecture

When I became a member of the NU faculty I made good use of this lecture by Bailar to prepare a fifty minute lecture which I gave to my freshman class each year. This I did primarily because, of the 600 or so in my lectures, only 5 to 10 might go on to become professional chemists. Others would become doctors, lawyers, politicians, etc. It is important that *all* develop a basic scientific literacy as it becomes more and more important in this technological world. In addition, others could be involved in making decisions on the support of science for the good of our society. These decisions on what is required to do basic and applied science must be made by individuals who have at least a basic understanding of science.

The freshman class of 600 students were told that the following week one of the lectures would be the most important lecture that I would give. They were invited to bring their girlfriend, boyfriend, or parents. The lecture would be on how science works, using the early history of metal complexes. Furthermore, they were told that they need not attend if they did not want to, because they would not be held responsible for it on exams—there was always some applause. I told them that if they were to attend that they would become somewhat knowledgeable as to what science is all about and why it is important in order for the US to stay ahead of the rapid developments in our technological age. In reality, almost all of the students did attend the lecture, and a few of them even brought someone. I gave my usual lecture on this topic, somewhat watered down, but still with examples they could understand as to why basic research may take a long time to solve a given problem. Examples were given to show how fundamental science has made important contributions to the problems of medicine, agriculture, energy, etc. The students reacted very positively toward the lecture. Since I gave the lecture each of the 40 years I taught the course, it became known on campus by freshman students who had the course, and a few would even show up to hear it a second time.

Figure 1-12. Sophus Mads Jørgensen (1837-1914), Professor at the University of Copenhagen, was an excellent experimentalist who prepared many of the metal complexes that contributed to Werner's formulating his coordination chemistry theory.

Most of the students were excited about research history and how science works and believed that they had a better feeling about science. I was also able to show them metal complexes prepared more than a century ago in the laboratories of S. M. Jørgensen (**Fig. 1-12**) and of Alfred Werner (**Fig. 1-13**). Jørgensen had prepared many such compounds, but his bonding theory was shown to be wrong. Because of the success of this lecture to students in my beginning class, I prepared a lecture entitled "How Science Works. Early History of Metal Complexes." The lecture was a big success, largely because it became known by many chemists. As a result, I was invited to give the lecture worldwide at universities and high schools. Slight changes had to be made in the lecture for these two audiences, but it was well received by both. The lecture was even given to groups with Nobel Prize laureates in attendance.

During my sabbatical in Denmark (see *Chapter 2*) I did research in the same laboratory that had been used by Jørgensen more than half a century earlier. In the lab there was a large glass cupboard containing the many new metal complexes that he had prepared and analyzed himself. I asked Professor Jannik Bjerrum (my host for the year) if I could take a couple of the compounds to show students. He said, "Yes, but always remember to tell them that they were prepared for the first time in Copenhagen, because most people think that Denmark only produces excellent butter, cheese, and ham." I asked for and was given *cis* and *trans* [Coen$_2$Cl$_2$]Cl (see pp. 21-22 for structures and nomenclature). I then bought a small Danish flag (red with a white cross in its center reaching to the four ends of the flag) and made certain I always showed it and told this story about Jannik in my lectures.

Arturo Miolati and Alfred Werner

Figure 1-13. Professor Alfred Werner, (1866-1919) at the ETH, gave us the coordination theory in 1893, after a century had been spent by some chemists trying to understand the nature of metal complexes. About the only experimental test of the theory available at the time was the measurement of the electrical conductivity of solutions of the metal complexes. This was done by his friend Professor Arturo Miolati (1869-1956) at the University of Padova, Italy.

My lecture given to the freshman chemistry class was often attended by some of our graduate students and postdoctorates. A few weeks after I gave one of these lectures, one of my postdoctorates from the University of Zurich came to see me in my office. He said, "I have a present for you," and he handed me a vial containing a metal complex. He explained that he had attended my lecture and saw that I showed samples of compounds made in Denmark and their flag, but none made from Werner's lab in

Switzerland and its flag. Therefore, he had asked one of his friends doing graduate work at the University of Zurich to go up to the attic in the late night hours and get a sample of one of Werner's compounds to send to him to give to me. He also sent a Swiss flag which differs ever so little from the Danish flag, the Swiss flag also being red but with a white cross in the center which does not reach out to the four ends of the flag. Later when I gave the lecture, I showed both samples and also the flags from the two countries. I did tell my friend Professor G. Schwarzenbach (see *Chapter 5*) at the Technical University of Zurich that I had been given this stolen compound, and offered to return it. He told me that I could keep it. After hearing the story about Jannik in Denmark, Schwarzenbach said, "When you show the compound and the flag, make certain you always tell them that the compounds were made first in Switzerland, because everyone thinks of it only as a place to ski."

When on a lecture tour of Switzerland universities, I gave a lecture at Eidgenossische Technische Hochschule (ETH) where Werner had been professor. I had given the Swiss university a choice of three lectures, two on our research and one on the history of metal complexes. It surprised me when Werner's university faculty chose the history lecture. My host showed me Werner's lab, the compounds that were made in his lab (giving me a sample of compound $[(NH_3)_4Co-HOOH-Co(NH_3)_4](SO_4)_2$, had me sit in Werner's chair, took me to the house where Werner had lived, and showed me his study. I also was shown the tavern Werner visited all too often, and, finally, visited his grave. Werner was the first person in Switzerland to get a Nobel Prize (1913).

In 1986, Harry Gray received the *Linus Pauling Award*. Three lectures are given during the day to celebrate the Award. One is the lecture of the Awardee, and the other two lectures are of his choice. Harry had heard me give my lecture on the early history of metal complexes and had seen the display of the century-old metal complexes made by Jørgensen and Werner, so Harry insisted that I give this talk. Linus Pauling was sitting in the front row with his characteristic sparkling eyes, listening intently. During the talk, I looked at him and wondered what he thought of my comments on how science works. After the talk, Linus came up to shake my hand and say how much he enjoyed my lecture. He had a good look at the century-old samples and told me he had read some of Werner's papers when he was writing his book *Nature of the Chemical Bond*. (In my opinion, this is the best book ever written on chemistry because of the many predictions made by Pauling. This prompted a large number of chemists to do research to test his proposals.)

In 1984, I was invited to give the opening talk at the annual *Peter A. Leermakers Symposium*, followed by Roald Hoffmann (Nobel Prize 1981). The title of my lecture was "Metal Complexes from Werner to Now," and Hoffmann's lecture was "Building Bridges Between Inorganic and Organic Chemistry." At the coffee break, he and I had a conversation about our lectures, and my recollection is that we agreed to disagree on a few points. The next time I gave this lecture was when I received the *Harry and Carol Mosher Award*. My good friend Professor Henry Taube (Nobel Prize 1983) attended my talk. Afterwards he enthusiastically told me how much he enjoyed the lecture for, as he said, it is very seldom one hears a chemist give a talk about anything other than his research. He then asked if he could make copies of my slides because he was going to Japan to give a few lectures, and they had requested that one should be of a more generalized nature. He felt, as he prepared this lecture, that he could possibly use some of my slides.

My Dissertation Research

The research problem that I worked on with Bailar had to do with the synthesis, characterization, and reactions of cis-[Pten$_2$Cl$_2$]Cl$_2$, where en is H$_2$NCH$_2$CH$_2$NH$_2$. The *trans* isomer had been reported by the Russians, but it seemed that there had been no attempt to prepare the *cis* isomer. It should be mentioned that there are large deposits of minerals in Russia that contain platinum, and, therefore, much of the early research on platinum was done by Russian chemists. They found this work so important that an Institute of Platinum Chemistry was formed.

Our interest was to obtain the *cis* form which would be optically active and allow us to investigate stereochemical changes that take place during ligand substitution reactions. After several different attempts, we were not able to isolate cis-[Pten$_2$Cl$_2$]Cl$_2$. The results of our experiments were published and the research done was the basis of my Ph.D. dissertation. Bailar was very generous in suggesting that I finish in three years, because he did not want me to get drafted in World War II and then have to return to get (or not get) my degree.

A previous graduate student, Betty Rapp Tarr, had worked on this problem for her Ph.D. dissertation. She also had not succeeded in obtaining the desired *cis* isomer. Worthy of note is the fact that Betty, at the time, was an exception, being one of the very few women pursuing an advanced degree in the sciences or engineering. By contrast, there are now many women doing research in these areas and making outstanding contributions to the field.

Now, before discussing my dissertation research and other work done during my career in chemistry, it would help to define some terms to be used and illustrate some of the structures involved. The generic term for the compounds is *metal complexes* because, as mentioned earlier, this was the term used during the 100 years when the nature of the compounds was not understood. The metal of the complex occupies the center of the structure and it is attached to surrounding groups called *ligands* (**1, 2**). The number of ligands attached to the metal is said to be its *coordination number*. Metals with a coordination number of six almost always have an octahedral structure (**1, 3, 4**). Metals with a coordination number of four can be either square planar (**2, 5, 6**) or tetrahedral (**2a**). Ligands in adjacent positions are called *cis* (**3, 5**) but, if in opposite positions, are called *trans* (**4, 6**). Some ligands attach themselves to more than one coordination site on the metal, and are called multidentate ligands (**8**). For example, ammonia (NH$_3$) is monodentate (**7**), since it occupies only one coordination site on the metal. Whereas, ethylene diamine (NH$_2$CH$_2$CH$_2$NH$_2$, given the symbol en) is bidentate for it is attached to two coordination sites of the metal. Thus, en is equivalent to two NH$_3$s.

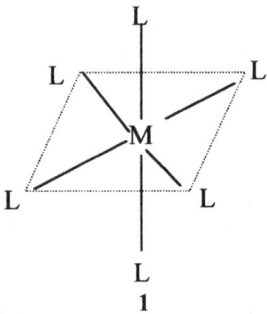

1

six-coordinate, octahedral
Solid lines represent chemical bonding.
Dashed lines are to show structure.

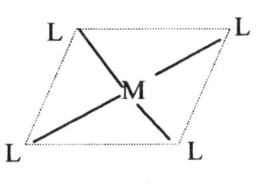

2
four-coordinate, square planar

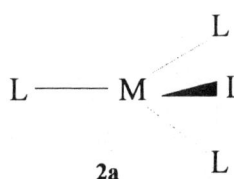

2a

four-coordinate or tetrahedral
no *cis-trans* isomers

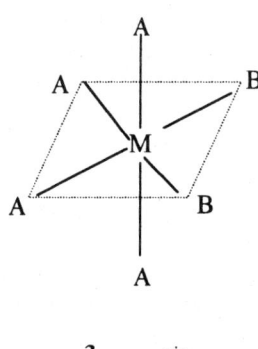

3 *cis*

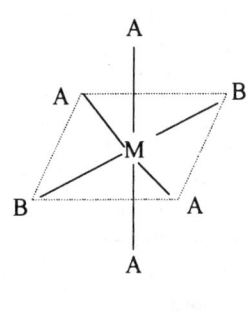

4 *trans*

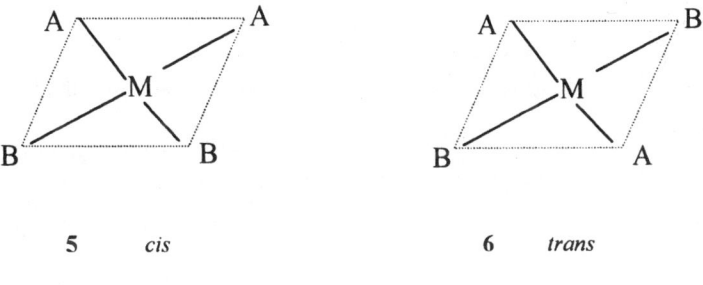

5 cis   **6** trans

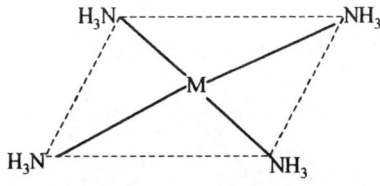

7 monodentate ligands

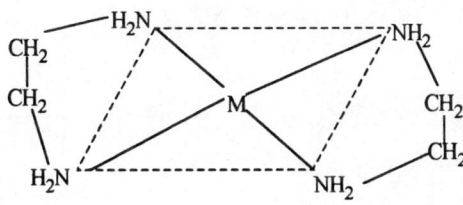

8 bidentate ligands

Back to my thesis research, which was to synthesize cis-[Pt en$_2$Cl$_2$]Cl$_2$, **11**, where en is $H_2NCH_2CH_2NH_2$. First, it was necessary to convert expensive platinum into the required starting material, H$_2$PtCl$_4$. This was done by the reaction of platinum metal with aqua regia (HNO$_3$+HCl), followed by the reduction of Pt(IV) to Pt(II), (eq. 1).

$$\text{Pt} \xrightarrow[\text{HCl}]{\text{HNO}_3} \text{H}_2\text{PtCl}_6 \xrightarrow{\text{SO}_2} \text{H}_2\text{PtCl}_4 \quad\quad (1)$$

At that time there was little or no funding, so one had to be very frugal with the use of precious metals. Therefore, all residues generated by the reactions involved in the research were saved and the Pt recovered. This was time consuming, but did permit the use of small amounts of Pt to carry out many reactions, and did provide students the experience of doing more syntheses.

Having prepared H_2PtCl_4, I then made several attempts to obtain the desired *cis* [Pt en$_2$Cl$_2$]Cl$_2$, **11**.

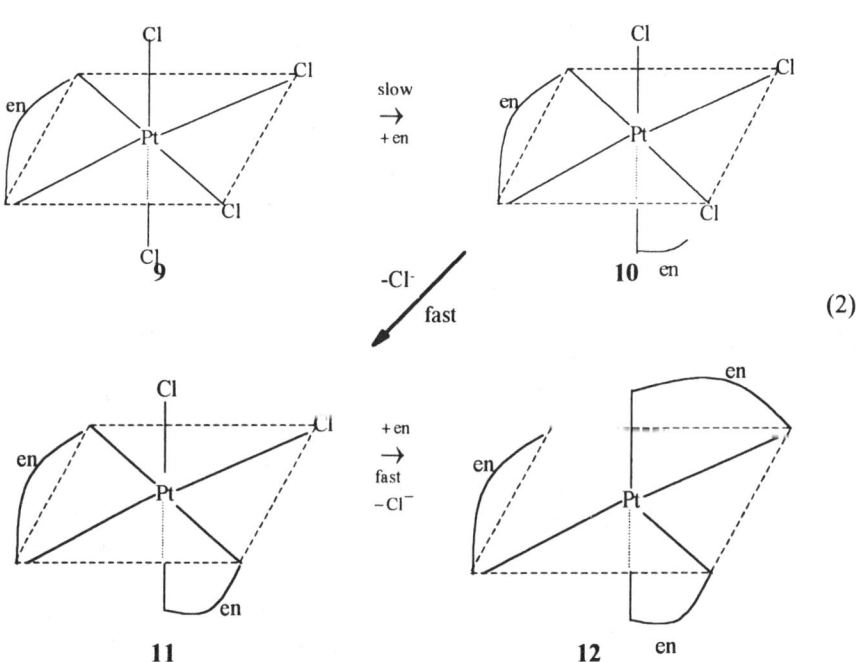

(2)

The reactions shown (2) were expected to occur as represented in terms of the Pt coordination chemistry known at the time. This was based on the discovery, by Chernyaev in 1926, that a negative ligand labilizes, i.e., makes more susceptible to substitution, the ligand in its *trans* position than the neutral ligands. If that is so, it should follow that the Cl ligand *trans* to the Cl$^-$ (**9**) would be more reactive than the *cis* Cl$^-$ ligands. Thus, **9** is expected to give **10**, and then, due to a favorable entropy effect, the monocoordinated $H_2NCH_2CH_2NH_2$ should readily close to give the *cis* bidentate ring structure **11**. (Entropy is a measure of disorder and, as disorder increases, the system becomes more stable.) If this happened as planned, it should yield the desired *cis*-[Pt en$_2$Cl$_2$]Cl$_2$, **11**. Unfortunately, I was not able to isolate either **10** or **11**. This can be explained because **9** is only sparingly soluble, so that it slowly reacts with en to form **10**, which then reacts rapidly to form **11**, which in turn gives the final undesired product **12**. All of the many attempts to vary conditions to obtain the desired **11** failed. Keep in mind that research in the early 1940s was not privileged to have all of the sophisticated instruments now available to do modern inorganic chemistry (such as NMR, infrared, fast reaction instruments, etc.) Because of these limitations, it was not possible to examine the reaction mixture for the possible formation of **10** and **11**. *cis*-Platin was not present. It was our target compound (**11**) that we failed to make. The three examples of instruments show the variety of instruments that later became available. Such apparatus which helps to study fast (femtoseconds) reactions would have been helpful.

The publication of these results in the *Journal of the American Chemical Society* attracted little or no attention, and there were very few requests for reprints of the paper. This was understandable for, as mentioned earlier, there was little interest in inorganic chemistry research in the US. You can imagine how surprised I was some twenty years later when I started to get requests for reprints of this paper from laboratories worldwide, and telephone calls asking me questions about our work on Pt chemistry. I then learned that Professor Barnett Rosenberg at Michigan State University had discovered that *cis*-[Pt(NH$_3$)$_2$Cl$_2$], **13**, has anticancer activity. This simple compound and its inactive *trans* isomer had been prepared about a century ago. Here, again, we see how fundamental research may lead to unexpected important commercial products. This *cis* compound represents a new type of an antitumor drug called *cis*-Platin, **13**. *cis*-Platin is known to be highly effective in the treatment of testicular, ovarian, bladder, head, and neck cancer. Almost complete remission of testicular cancer is obtained by this drug. The interest in my thesis research was because I had been able to prepare [Pt en Cl$_2$], **14**, in good yield, which is closely related to *cis*-Platin.

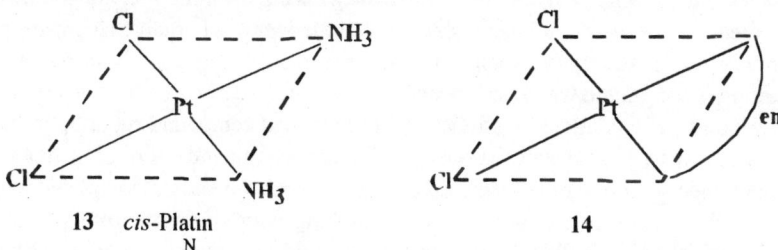

13   *cis*-Platin
N

14

Because of this close relation between **14** and **13**, where en almost amounts to 2 NH$_3$, investigators wanted to test the anti-tumor activity of **14** versus **13**. Many chemists prepared compounds similar to **13** using different metals and/or different ligands to be tested for their anticancer activity. Some of the new compounds tested also had anti-tumor properties, but none were much better than that first reported.

The discovery of Rosenberg is a good example of how serendipity plays such an important role in science. Rosenberg was doing experiments using Pt electrodes to carry out the electrolysis of a solution to see what effect it might have on a certain bacteria. A most unusual effect was observed which had not been expected. He wanted to know what caused this, and he was able to determine it was not caused by the electrolysis but by the trace amount of Pt in solution coming from the Pt electrodes. Further experiments were carried out using known Pt compounds which finally led to the discovery of the anticancer properties of *cis*-patin. Serendipity results when some unexpected behavior is made in the laboratory and is then intelligently recognized as perhaps being important enough to investigate in detail. Some people refer to this as an experimental accident, but it is much more. Also these "accidents" are often some of the most important breakthroughs in science. That is because other research usually involves extending work already understood. Serendipity involves observing experimental results that no one could have predicted, and opens up new areas of science. As stated by Pasteur, "experimental accidents happen more often with the intelligent investigators."

Finally, of interest to me was the failure of the Cancer Institute of the National Institutes of Health to find the antitumor activity of *cis*-[Pt(NH$_3$)$_2$Cl$_2$] in their tests of many research compounds. These tests were carried out several years before the discovery of *cis*-Platin, when *all* commercially available compounds and samples of research compounds were systematically tested for anticancer properties and for side effects of the compounds. During my early years on the faculty at NU (to be discussed later), we had prepared several Pt compounds in order to study the kinetics and mechanisms of substitution reactions. Among the many metal complexes I provided for this screening was *cis*-[Pt(NH$_3$)$_2$Cl$_2$], which did not exhibit any promising anticancer activity. The Cancer Institute tests drugs by using mice with cancer to determine if the compound being tested cures the cancer. Low concentrations of the tested compound is necessary, because many anti-cancer compounds are toxic. Most of our compounds tested were not of interest because of high toxicity and, as a result, many mice were killed.

In addition to my interaction with Professor Bailar and my dissertation research, here are a few things which I vividly recall happening during my three years in graduate school. I mentioned earlier that I had learned what little I know of organic chemistry in a qualitative organic laboratory course. At that time, in such a course, about the only way to identify a compound was to isolate one (or more) of its derivatives and determine its melting point. If this agreed with that of the presumed compound reported in the literature, there was a good chance the two were the same compound. Further assurance could be obtained by mixing the unknown compound with the known compound and taking the melting point of the mixture. If the melting point of the mixture was the same as that of the individual two compounds, one could reasonably conclude that the compounds were the same, and the original unknown had been identified.

This brings to mind a frightening experience which I shall always remember. The derivative of the compound that I was given to identify required the use of toxic mercuric chloride (HgCl$_2$) and a slightly elevated temperature of the reaction solution. This was done in a stoppered test tube that I put in hot water. After a short time, when I believed the reaction to be complete, I removed the stopper in order to isolate the expected product. Removal of the stopper resulted in a drop of pressure of the reaction mixture causing an explosive-like escape of the solution directly into my face. I was not certain if my mouth was closed when this happened, and I began to worry that I may have swallowed some of the toxic mercuric chloride solution. What I did was stupid, so I did not want to tell my instructor about it. I did seem to recall that raw eggs were a good antidote for mercury poisoning. I decided to go to an ice cream place near our chemistry building and asked them to give me a milk shake containing one raw egg. I had a second one, and they looked at me as if I were crazy. Not knowing what else I could do, I went to my room and lay in bed to try to take a nap, not knowing if I would die and not ever awaken. When I did awake, I had a good look around to make certain I was in my room and not in heaven. I will never know if my mouth was closed when the reaction solution hit me in the face or if the milk shakes spiked with raw eggs were responsible for my being here to write this autobiography.

One extracurricular activity that I vividly recall was perhaps largely the result of the six of us inorganic chemistry graduate students being vastly outnumbered in the total of more than two hundred graduate students. For this reason, being the minority group, we were looked upon as students who should be given meanly duties in the department. During my first year of graduate work, I lived in a room in a private home. My second year I joined the Alpha Chi Sigma fraternity and moved into their house. This house was occupied by about thirty students. It had been purchased some years earlier by the department of chemistry and was slowly being bought by the fraternity with income from room rents and gifts. The house had previously been a sorority house, as was easily noticed in the washrooms, which were devoid of urinals.

Living in the Alpha Chi Sigma House during my last two years afforded me the opportunity to get to know most of the graduate students, other than just the six inorganic students. This applied mostly to the thirty students who lived in the house, but also to many of the other students invited to attend several of the house activities and parties. Some of the many friends I made during this time have had illustrious careers in their subdisciplines, and a few of us still maintain contact.

My last year of graduate work I was elected "Master Alchemist." This meant that I was responsible for all of the things having to do with the house, such as arranging for needed repairs, making certain the quality of the food was OK, assuring the house was clean and in order for receptions and other activities, etc. Faculty were invited to two functions a year and several of them would attend. If Professor Roger Adams was in town, he would come. Since he was chairman of the department and a top organic chemist, his attendance at one of our activities meant that more of the faculty would be present. These functions were informal and gave students a chance to interact with some of the giants of chemistry. There was a pool table in the house, which made it possible to enjoy playing pool with faculty. The important feature of these gatherings was the relaxed interactions with world renowned faculty, and seeing that they, too, are only human, and not rigid know-it-all professors.

On one occasion, Roger Adams had just returned from Washington DC, where he often had to meet with committees on matters relating to World War II. He had taught an organic chemistry course that semester and had given the students their final exam. As a few of us were talking trivia with Adams, one of the students who had taken his course asked if he had found time to grade the final exam. Adams said that he had and that the student had done very well. The student then asked if he had read carefully his answer to question three and Adams replied that he had, at which time the student handed him a cigar. Adams inquired, "Why the cigar?" and the student told him that at the end of question three he had offered to give him a cigar if he had, in fact, read his answer carefully. Adams responded that the student's answers to the first two questions were so good that he had not thought it necessary to read the remaining two.

One other amusing event happened on a weekend when our homecoming football game was played against IU. It was traditional for all the fraternities and sororities to decorate their houses and front lawns for homecoming weekend. The sorority across the street from our house put up very nice decorations, including a poster which announced, "We are laying for Indiana." They had not realized the various interpretations of this statement until a few of us called and said, "I'm from Indiana—and I'll be right over." After this, they immediately removed the sign.

I found that graduate school was not only learning more chemistry and how to do research, but, in my case, it made me realize that I, too, one day might be able to contribute to the profession as was being done by the chemistry faculty at the UI. For me, the three years spent in graduate school were also a growing up process, where I developed a sense of responsibility—something that was missing when I arrived at the UI. I owe most of this to John Bailar, my mentor and good friend. His colleagues on the chemistry faculty were also very helpful, as were my interactions with many of the graduate students.

INDUSTRIAL RESEARCH—ROHM & HAAS (R&H)

As mentioned earlier, Professor Bailar encouraged his students to get their Ph.D. as quickly as possible, in order to avoid any interruption in their careers due to World War II. Largely because of this, I obtained my doctorate degree in three years, at age 23 in 1943.

World War II was strongly supported by Americans, and several of my friends had already been drafted into the army as GIs. However, the government had classified some research programs in progress as being more important than its need for army personnel. I decided, if that was the case, then I would take a position doing classified research of interest to the war effort. I was invited to visit the University of Chicago, and its Manhattan Project, and the R&H Chemical Company in Philadelphia. Since the research I would be involved with was classified, neither of these two places was able to tell me specifically about the research that I would be doing After much consideration of the two offers, I finally decided to accept the R&H offer. For the most part, my decision was made on the basis that I was familiar with some of their unclassified research, i.e., ion exchangers and plexiglass. Plexiglass is a clear, glass-like polymer which was being molded to fit around the noses of airplanes so that pilots would have maximum vision. Ion exchangers are used in a variety of ways, including the softening of hard water. The exchanger can be treated for recycling. I thought that R&H would enable me to do classified research that the army wanted done and, in addition, when the war was over and there would no longer be need for such work, I would have a good chance to get appointed to a permanent position with the company. I still think, in retrospect, that this was the better choice for me.

However, I know several chemists who did important and challenging research on the Manhattan Project. They did some of the seminal research on the chemistry of metals, in particular, the transition metals, the lanthanides, and the actinides. Their research, at the time and later in academic or in government laboratories, contributed immensely to the serious beginning of research on inorganic chemistry in the US.*

At R&H, my research director was Loran Hurd, who had obtained his doctorate at the University of Wisconsin. He was a good guy—energetic and willing to let us work almost independent of him on our research. One other older person in the group was a German-American who was said to know more about zirconium (Zr) chemistry than any other person in the world. The veracity of this is not known, for he was a loner who spent most of the time in the library, perhaps keeping up with the Zr literature. In addition, besides myself, there were two young Ph.D.s from MIT and four assistants. One of the assistants had worked in the lab for many years and knew more than we did about where to obtain what we needed for our research. He was always willing and ready to help, and I certainly made good use of him.

Initially, I was assigned the task of preparing several compounds of Zr to be tested for fire retardant properties. I should mention that all of the research classified at that time is now unclassified. Although I prepared and characterized several Zr compounds, none of them, unfortunately, reached the standard required for a fire retardant compound. The compounds were also tested for water repellant properties and some were good enough to compete with existing commercial products presently in use but, to my knowledge, none of these Zr compounds are being used for this purpose.

The interest that R&H has in Zr chemistry is because it has access to large deposits of Zr minerals. Their largest commercial product is ZrO_2, which is white and opaque, as is TiO_2. The two products compete in the paint and enamel markets. It is said that another outlet for Zr is in tanning leather for use in baseballs. Zr produces a a leather that is white throughout, so that, when a ball is scuffed, it continues to be white rather than having a gray scuff mark.

My next project was to prepare a mica substitute, and I am pleased to say that this was successful. The product was enough like mica that, had the army needed it, it could have been used in place of mica. After several empirical attempts, the best results were obtained by using montmorillonite impregnated with a mixture of phenol and formaldehyde. This was then formed into a film which was heated to a temperature that caused

---

*Inorganic chemistry has now become one of the most important areas of research. It encompasses research on all of the elements in the periodic table. For example, elegant significant work is taking place in the areas of main group elements, solid state inorganic chemistry, new materials, organometallic chemistry, and bioinorganic chemistry. No longer is it considered a second class subdiscipline. It now challenges organic chemistry for first place, as more and more graduate students choose to get their doctorate degree in inorganic chemistry. Chemists of my generation are fortunate in having seen inorganic research in the US grow to its present important position. I was particularly pleased a few years ago when the Italian Academy of Science Lincei, of which I am a foreign member, asked me to give a talk on "Frontiers of Inorganic Chemistry." Later, I was invited to write an article on this subject, and I accepted. This afforded me an opportunity to scan the scientific literature and see how far inorganic chemistry has come since my graduate student days. I am delighted and gratified to have lived to see inorganic chemistry rise to its present status in my lifetime.

the phenol and formaldehyde to react, forming a resin inside the host montmorillonite. The process was developed to a pilot plant state to produce quantities of mica-like films. Of the various problems we encountered, the most difficult was that of having to carry out the process in a dust-free laboratory because any dust particle would create a weak spot in the film, causing it to fail in its use as a mica substitute. This required the use of a special dust-free room and all workers had to wear white dust-free clothing over their regular clothes. The use of dust-free research and work is now often used, but then it was a bit unusual to have workers dressed for a special environment.

In retrospect, the best time for me at R&H was that spent in the library. I was able to do this because I had a very good lab assistant, who could reliably do much of our laboratory work. In the library, I read most of the papers published by American inorganic chemists. This did not require much time, because there were only a dozen or so chemists in the States doing research in inorganic chemistry. Furthermore, very little of what was being reported by these chemists was of interest to me. As a result, I spent much of my time browsing through the *Journal of the American Chemical Society* (*JACS*). One thing that caught my attention was that organic chemists were not just using known named reactions to synthesize compounds, as I was taught in graduate school. Instead, I read several exciting and interesting articles that dealt with details on how the reactions took place. This involved a study of the kinetics of the reaction which, in turn, was used to propose a mechanism of how the reaction takes place. In particular, the one thing that attracted my attention the most was a mechanism known as either an S_N1 or an S_N2. Both reaction types are nucleophilic substitution reactions which, in organic chemistry, means the replacement of one group by another, almost entirely on tetrahedral carbon. Later, when I discuss my arrival at NU, it will be seen how my browsing in scientific journals of the 1940s in the R&H library helped shape my career in chemistry.

There are things other than chemistry which I recall from my three years at R&H in Philadelphia. One thing that still troubles me, at times, is the fact that I chose to do government classified research instead of joining the army. My only redeeming feature was that I felt this decision was best made by the military. If they believed that my doing research on one of their classified projects was more important than their having one more GI in battle, I decided that I should do what was asked of me. We all know that World War II was what had to be done, and that it had strong support of the people. That being true, it was not easy for a healthy young man to have to be seen in civilian dress. For instance, I had to go home twice during this period because, on both occasions, the doctors believed my father would die of cancer in a couple of days. They were correct the second time, in 1944. In order to get home, I had to take a train from Philadelphia to Chicago and then change to another train to Southern Illinois. I remember that the train to and from Chicago was completely full with persons sitting on their luggage. In our car, I was the only male about my age not in military uniform. There were mothers who had lost their sons in the war, younger ladies who had lost their husbands, and young ladies who had lost their boyfriends. I had the feeling that they and the men in uniform were all looking at me and thinking I surely must be a "draft dodger." One woman I sat next to did ask me what ailment I had that kept me out of the army. I told her that I was healthy and could be in the army, but was deferred because the army always felt it was more important that I stay in the lab and continue doing research on their classified problems. She seemed satisfied, but I felt that all the others on the train were of the strong opinion that I should be in uniform.

Some fifteen years later, when we were all back in civilian clothing and little more was being said about the war, our two small children asked my wife and me about the war. They had just come in from playing with their friends next door and, at the dinner table, they told us that Katy said her father was in the army and went to Italy to fight in the war. Our children then asked me, "Where did you go to fight, Daddy"? My wife and I had to try to explain to

Figure 1-14. Our R&H baseball team in the city of Philadelphia minor league in 1943-46.

them that I did not go to war, but stayed in Philadelphia and did research for the army. We told them the army felt this was more important than having me go to war. They seemed to understand. However, it was clear to my wife and me that they would rather have had me go to war so that they could tell their friends where I went to fight.

When I began working for R&H, it was still a family company. Its plant was located in Bridesburg, a small suburb of Philadelphia which is along the Delaware River and was inhabited largely by Polish-Americans. The research lab was at the same site as the plant, so all of us had to punch the clock when we entered or left work. We were told that this was done so as to not upset the plant employees who had to keep track of the hours they worked. We were never certain that this was not also done to keep an eye on the work ethics of the Ph.D.s doing research.

On Fridays, the lunch hour was somewhat longer than usual, for there were several bars near the plant that both plant and lab personnel would frequent. The bars had dart and shuffle boards which were used to compete for drinks. The drinks might at times be a "shot and a beer" which is a whisky and a beer chaser. The food was mostly excellent Polish sausage. This rather long lunch hour meant that little work was accomplished in the afternoons.

Behind the research lab there was a rather large open area and some of us, at times, would not have lunch at the bar. We would, instead, have a quick cafeteria lunch, then go out in back of the lab and play around a bit with a baseball. In fact, during the baseball season, we had an R&H team in the city league. None of us were very good, but since most good players were in the military, we did manage to make the team and to occasionally win a game. I played second and batted seventh or eighth, telling you something of my baseball talent (**Fig. 1-14**).

One other thing I recollect about my three year stay in Philadelphia was how four or five of us who were not married would eat out together once a month. We always tried to pick a different restaurant, but always one reputed to be "one of the very best" in Philadelphia. The most outstanding place was the old original "Bookbinders." It is "world renowned" for its sea food. Having been in the state of Illinois all my life prior to this, I had never tasted sea food. All I had ever eaten were the perch, catfish, and bass that we caught in small ponds or small "rivers" called "creeks" in Southern Illinois. My colleagues, who enjoyed sea food, always chose appetizers of a half dozen raw blue point oysters or cherrystone clams on the half shell. It made me ill watching them eat and enjoy these slimy things. However, I decided that if my friends liked them so much, I should close my eyes and try them. I started by ordering a half dozen cherrystone clams. I succeeded in eating only one and giving the other five to my friends. The next time I managed to eat four, etc., until now I have developed a real gourmet liking for all kinds of sea food. Another restaurant we liked to frequent was "Arthur's Steak House," for it always had the best cut of steak in Philly. However, at times the restaurant would be closed with a sign stating, "Closed until we get our meat stamps." This was during World War II, when meat was rationed. The steak house, wanting to maintain its reputation for serving the best and largest steaks, would wait to open until it had enough food stamps in order to get the meat required to meet their high standards.

Two amusing things I recall about R&H had to do with two of its products: plexiglass and a leather tanning agent. Plexiglass is polymethylmethacrylate, at the time used in large quantities almost entirely for the front nose area of military planes, as mentioned above. The people at R&H were always thinking of other uses for plexiglass, since the exhaustive use required by the army would not last much beyond the war's end. One of these other applications was to make a round mold of plexiglass to fit over the top front half of a coffin, with a transparent cloth over the plexiglass, through which one could see the deceased. If one wanted a better look, this was possible by just brushing the cloth back and viewing through the plexiglass. This seemed to be a good outlet for the product, until a disturbing event happened. After several brushings of the covering cloth over the plexiglass, static electricity developed. This resulted in a mourner, who had pushed the cloth aside, experiencing a sudden shock when the hair of the deceased jumped upright.

Finally, a comment about how R&H got started. Although the details of this are not known to me, here is what I have been told. Rohm was a Ph.D. chemist and Haas was a businessman with the needed funds to launch such a venture. For some reason, Rohm had gathered information on the use of dog dung for the tanning of leather. He was of the opinion that certain enzymes in the dung were responsible for the successful treatment of leather. They later came to the US and started a small family company. Several years ago stocks were issued, and today it is a chemical company with earnings of billions of dollars a year. It can be said that here is a large corporation which got its start with dog dung.

2

MARY, THE CHILDREN, AND ME

Having had two years of chemistry, I was assigned to work in the student health laboratory during my junior and senior years. My job was primarily to do urine analyses and hemoglobin tests. There I met Mary Nutely, a work-study student member of the staff in the office of the university doctor. She would often bring me samples in the lab to test, and we soon became good friends. Mary was taking a freshman chemistry course and told me that she was having trouble with it. I offered to help her and thereafter we began to be with each other more often. In spite of my help she failed the course, but she never held it against me.

Her chemistry instructor was Professor Kenneth Van Lente. He became good friends with Mary because Ken knew her Aunt Liz, who was also a SIN faculty member. For many years Ken would always ask me about Mary, and tell me she was a fine lady and I should take good care of her.

Before coming to SIN, Mary lived with her mother and siblings in Long Meadow, Massachusetts, a suburb of Springfield. The children were all young when their father died of a heart attack. Their mother had a difficult time financially, but was able to supplement her small income by playing the organ for weddings and other functions at the Catholic Church. The children went to work as soon as they were old enough, for it was necessary that they contribute to the family finances. Mary was the only child to go to college and, of all places, went to SIN, which was unknown to persons in Massachusetts and such a long distance from there. She was often asked about this, and would have to tell the story of how it happened.

Mary came into my life at SIN because her Aunt Elizabeth, the sister of Mary's mother, was left alone with three young children after the sudden death of her husband.

Aunt Elizabeth worked full time at SIN and needed help with the children and other household chores. Mary's mother and Liz decided that Mary could go to Carbondale to help her aunt and, while there, she would attend SIN. Mary did this, and registered for a light load of courses in order to be able to be of more help to her aunt.

Mary and I were very young when we became good friends. We would spend time talking about our families. She told me about her mother's difficulty staying within the monthly budget, and I told her about my family's plight in Coello. I do not believe that we ever thought of ourselves as girlfriend/boyfriend. Mary only stayed two years at SIN and never did finish college. She left because she had to return home, get a job, and make her contribution to the family. During my remaining year before getting my B.Ed., she and I wrote to each other. However, this correspondence stopped shortly after I began graduate school and as Mary began to have other interests at home.

Upon obtaining my Ph.D. in 1943, I took my first job as a research chemist at the R&H Chemical Company in Philadelphia. After being there for a few months, I realized that this was not far from Springfield, where Mary lived. I proceeded to procure the train schedule for the Philadelphia/Springfield, Massachusetts circuit. I found that the train ride would be convenient and OK timewise with my work responsibilities. I therefore wrote to Mary, stating that I had finished my graduate studies and had taken a job at R&H Chemical Company in Philadelphia, which was not too far from Springfield. I did not mention coming to visit her on some long weekend because I did not know if she had a boyfriend or might even be married, for it had been three years since our exchange of letters. Fortunately, neither of these had happened, and she wrote to me immediately. She told me she had often thought of our friendship at SIN and wondered if we would ever meet again. The two of us then took full advantage of the train rides, as I would often go to Springfield to see her and she, at times, would come to Philadelphia.

Three years later (1946) I accepted a position as chemistry instructor at Northwestern University in Evanston, Illinois, a suburb of Chicago. Before my leaving Philadelphia, Mary and I were engaged, with the understanding that we would get married after my first year at NU. This happened, as planned, on 14 June 1947, in a small Catholic Church in Longmeadow. Mary's mother played the organ, after which we had a very nice sit-down lunch followed by a high-spirited Irish-Italian celebration. The rest is history, but I will tell it, as I know Mary would want it told. I would like to say that had it not been for Mary (**Fig. 2-1**), this autobiography would end here.

After our wedding, we went to Montreal for our honeymoon. One thing I will always remember is that when Mary became aware of my interest in the U.S.

Figure 2-1. Mary and I celebrate our first Christmas after our marriage on 14 June 1947.

Open Golf Tournament, she insisted that I watch the final round on TV. This convinced me that I had made the right choice in selecting the right woman as my wife—one who was thoughtful and devoted. Upon our return from Montreal, we went to her family home in Springfield to gather her belongings, buy a car, and say good-by to her family. This departure was made easy because her family was so delighted to see us happily married. Mary's mother (**Fig. 2-2**) said that she felt as if she had gained a son rather than having lost a daughter.

HOME IN EVANSTON

We were both anxious to get to our first home in Evanston. We drove the 1000 miles with only one overnight stop. This was not bad, considering that there were no interstates at the time, in addition to the

Figure 2-2. Mary's mother in front of the church where we were married, and where she was the organist.

necessity of driving our new car slowly for its initial miles. Our first home was the attic apartment that I had shared with Bernie Adelson during my first year at NU. I had made the arrangements for our apartment before I had left to get married.

The chemistry department had a shortage of offices, so most of the new instructors had to share an office with another instructor. I was fortunate to get an office with Bernard (Bernie) H. Adelson, also a new instructor, who had just received his Ph.D. in organic chemistry with the late Professor Robert Baker. Bernie had also just been admitted to the NU medical school. He has since become Bernard H. Adelson, M.D., one of the very best doctors of general internal medicine, having received many awards for his outstanding work. Mary and I were always proud to be listed among his friends, and he also took good medical care of Mary and myself, a task which he continues to perform for me. I am certain that under his care a few more years have been added to my life.

Bernie, another graduate student (Arthur Schlesinger), and I shared an attic apartment near the university. When my roommates knew that I was to be married and would need to have a less expensive apartment, they said that they would move out so that Mary and I could move in. However, the lease on this apartment ran out after our one year. As fate would have it, NU was now making available low-rental temporary housing for faculty, consisting of a cluster of long military, round-top metal Quonset huts. These were located on campus, making it easy for faculty to get to their 8:00 AM classes. Mary and I lived in this temporary housing for the next couple of years. In the summer, the indoor quarters became very hot, and in the winter, at times extremely cold. The heat was provided by a gas stove to which one would go when very cold, but this meant that one became very hot on one side of the body and very cold on the opposite side! Another interesting feature of living in these huts was that they were divided in half with only a metal partition between the two parts. This meant that any noise or loud conversation that was made on one side of the partition was easily heard by your neighbors on the other side. After an evening out, upon returning home and switching on the lights, one could hear the cockroaches scrambling into the neighbor's dark quarters.

Our first child, Mary Catherine (MC), was born during our first year (1948) in this "temporary" housing, which, because of the duration of our stay, seemed as though it was fast becoming permanent. During the second year of our stay in these huts I kept badgering the staff in Human Resources to help me find suitable housing that we could afford. I was finally told that they had a house on Grant Street, about a 15 minute walk to the campus. This would be rented to us for not much more than we were paying for our hut. The house had three bedrooms upstairs which we rented to three of our chemistry graduate students. Our family occupied only the downstairs. The rent provided by these students was more than enough to pay our rent for the house. Since the residents of Evanston complained about NU having tax-free property not being used for educational purposes, the university did not want to own this house, which was off-campus. For this reason we were lucky to have the house offered to us at a much lower cost than its assessed value. Fortunately, as a faculty member, I was considered reliable and managed to get a mortgage that we were able to handle. Mary and I now had our very own home.

It was not long until our son, Fred Jr., was born (1950). This seemed like a reasonable family, with a girl and a boy. No plans were made to have two more children, but as you will find out shortly, we ended up having a total of four children: three girls and a boy.

Mary, by any standard, was the very best mother and the very best wife. Like most spouses in that generation, she did all of her work at home—taking care of the children and of me. She was a loving mother and devoted much of her time to the children. Now, by comparison, as I look back at the time I spent with the children, I realize how Mary was largely on her own with this responsibility. I need not tell this to chemists active in research and/or teaching; they get invited to give seminars at universities, lectures at scientific meetings, service on committees, and consultation for industries. These commitments require a lot of time—time that could better be spent with family. During the years when I was very active, Robert Burwell, one of my faculty colleagues, said, "Basolo, if you were a woman you would always be pregnant, because you have not learned to say no." Now, I see that he was correct and I realize that often this was at the expense of my family. In addition to taking care of the children, Mary also fixed our meals, did the housecleaning, paid the bills, and invested wisely the small amount of extra money we had from time to time. Yes, we chemists are fortunate to have "Marys" for wives, and we could not accomplish our achievements without them.

As MC and Freddie, Jr. were growing up, we, as a family, never had a real vacation. However, each summer Mary and the children would spend two weeks with her mother in Longmeadow and during their first week there, I would drive to New Hampton, New Hampshire, to attend the inorganic *Gordon Research Conference* (more later on GRC). After a week at the conference, I would travel to Longmeadow and we all would enjoy a good week together.

SABBATICAL YEAR IN DENMARK

When the children were 6 and 4 years old we went to Denmark for a year. I had received a Guggenheim Award, which made it possible for our family to spend a year in Copenhagen. This was my sabbatical year (1954-55) which I spent at the University of Denmark with the outstanding coordination chemist Professor Jannik Bjerrum (**Fig. 2-3**) as my host. We traveled to Copenhagen on the Osloford, a very fine Norwegian ship. The trip took seven days, the food and wine were excellent, and the sea was very calm much of the time. The passengers could keep occupied all day and evening with various activities that were available. Mary and the children kept busy playing games much of the time. My situation depended on my making good use of Dramamine. For example, the first night, we all went to see a movie and part way through the picture the ship started to move a

Figure 2-3. Professor Jannik Bjerrum (1909-92), at the University of Copenhagen. He is best known for his discovery of a method to determine the step-wise stability constants of metal complexes in aqueous solution by pH measurements.

little. The stage curtain began to sway slightly back and forth, back and forth, and I found myself watching that motion and becoming ill enough to have to leave.

When we docked, Jannik and a couple of his assistants were waiting for us at the port. They took us and our belongings to a house in a suburb of Copenhagen, a very nice residential area with houses built after World War II. The owners of the house had just left a week before to spend a year in the US. The lady was an x-ray crystallographer who would be at MIT, and her husband was a physicist who went to Harvard. They made things easy for us because the house contained everything that we needed, such as dishes, bedding, etc. They had left the key to the house with the neighbors next door, who were expecting us when we arrived. Our neighborhood welcoming committee consisted of three ladies who were waiting in the house with tea and goodies that they had prepared for us. They had also bought milk and other food for us so that we would be able to have the first day or two to simply adjust to our new surroundings. This gave us some extra time to learn how to shop in a different country. They said they were very pleased to have Americans nearby, for it would help them practice their English. Mary did not see a refrigerator, and upon inquiry was told that none was needed. The milk and other foods were kept outdoors in a small room attached to the house because the temperature in Denmark is cool enough that the food does not spoil.

Mary tried to learn some Danish but gave it up quickly after what happened one day as she went to catch the bus to the city. Their public transportation is almost always on time, and Mary arrived a minute or two late. A middle-aged man was standing at the bus stop waiting for the bus. Mary tried her Danish by asking "bussa a comma now?" He replied in almost perfect English, "Yes, lady, the bus is soon to arrive." After that experience, Mary never again tried to speak Danish. Except for the more elderly Danes, all of them spoke English. They taught English in schools as their second language. Before WWII and their occupation by the Germans, it was their second language, but this was stopped just after the Germans departed. Many of the Danes knew German but they absolutely refused to ever again speak the language. They told us about many unnecessary terrible things done by some of the Nazi soldiers. One thing we were told that happened involved their King, who felt he was just a Dane as were all other Danes, and the Danes liked him very much. At some point, the Germans requested that all the Jews in a small area of Copenhagen come out wearing the star of David, otherwise they would all be taken away. The problem was resolved when the King appeared wearing the star.

Winters in the Scandinavian countries are gloomy, with little daylight, and even on a "bright sunny day," the sun never gets overhead but shines directly into one's eyes. No doubt as to why these countries have the highest rate of suicides, with the winters taking their toll on the depressed. This was so noticeable to us that when Professor David Hume, on sabbatical from MIT, went back to the US for a week, the question most asked of him was, "How was the weather at home?" He said it was a pleasure having a few nice sunny days with the sun directly overhead.

One of the formal events attended shortly after we arrived in Copenhagen was a dinner honoring Professor Linus Pauling (1901-94) after he had just received the 1954 Nobel Prize in chemistry. This dinner, along with his lecture, was held at the University of Lund which is readily reached by ferry from Copenhagen. (Now the largest suspension bridge in the world spans this distance.) The dinner was a big success, and the talk by Pauling was informative but most enjoyable, as are his talks to mixed audiences. One thing that I was not prepared for happened during the dinner. I was seated directly across from a Swedish lady who kept staring at me, and finally said, "That is a very attractive tie that you are wearing, with its fine color and design." I thanked her and then wondered what kind of ties the other men would be wearing. The ties of almost all Scandinavians are of a formal solid dark color. Later, when I asked my Danish friends about this, they told me that they were more formal in dress and culture than Americans, and that the Swedes were the most formal of all.

The people living in this new residential area were about our ages, with children about the same ages as MC and Freddie. While we were having tea with the neighbors, Freddie went out to play with a next door neighbor's child, who was kicking a soccer ball. Freddie went to pick up the ball and was almost kicked in the face. This was his first experience that "football" in countries other than the United States really does mean *foot*ball, because no one is allowed to use their hands. After being there a week, one evening when I got home from the lab, Mary said, "Guess what happened today?" She

Figure 2-4. Our two children, MC and Freddie, in Denmark before the well-known mermaid in the Copenhagen harbor.

Figure 2-5. Our children, with their friends, waiting for the school bus to take them home. MC is the girl at the front left. Freddie is the boy slightly out of the picture at the right.

went on to say that when Freddie came in from playing with his friends, he said, "Mom, those kids sure talk funny." He had been playing all week, not even realizing he was in a foreign country!

It was clear from the beginning that the people living on our street were going to treat us as their guests. We were often invited for tea or dinner at their homes, and Mary made certain to return the same. This was a delightful year for all of us, as well as a very good year for me professionally at the university.

Mary and I knew about the famous Danish story writer of books for children, Hans Christian Andersen. Before our trip to Denmark we read his stories about "The Little Mermaid," "The Ugly Duckling," etc. to the children. Therefore, it came as no surprise that one of the first things the children wanted to see when we arrived in Denmark was the statue of the mermaid (**Fig. 2-4**), which is at the edge of the boat harbor in Copenhagen. What we found there was a small statue that was not at all what the children expected, and MC ended up in tears. Much the same thing happened on the birthday of the King. On that day the King and Queen came out on their balcony to thank the crowd of people who gathered to wish him a happy birthday. Again, MC and Freddie were shocked to see the King and Queen in their street clothes, and not the magnificent dress our children had expected.

The children quickly made friends (**Fig. 2-5**), and their new friends would come each morning to get them so they could wait together for the school bus. They went to the Danish school, and by Christmas (we arrived in September) they were able to communicate in Danish. Mary and I were not good in speaking the Danish language, but we did not really need it, for most Danes speak English. I recall one night at a nice, rather formal dinner at a professor's house that the host's large dog came to the dining room and was told in Danish to return to the kitchen. I looked at Mary and said, "Even dogs know Danish, but you and I do not." This was enough to cause loud laughter among the Danes.

The laboratory where Professor Arthur Adamson (on sabbatical from the University of Southern California) and I were to work had a great deal of history. It was the old lab used by Professor Sophus Mads Jørgensen to prepare many of the metal complexes that helped Alfred Werner formulate his coordination theory. The lab was likewise used by Professor J. N. Brønsted (1879-1947), well-known for his acid-base theory. The Danes told me Brønsted was to receive the Nobel Prize in 1934, but it had to be postponed and given to Professor Harold C. Urey (1893-1981) for his discovery of heavy hydrogen. If true, Brønsted was forgotten and never received the Nobel Prize (1935). Professor Niels Bjerrum,[*] father of our host had been a student of S. M. Jørgensen. When Niels learned of my interest in metal complexes, he, at times, stopped in the lab to tell me some stories. One that I recall was how Jørgensen was angered by Werner's attack of his chain theory, so much so that he refused to let

Figure 2-6. After dinner drinks and coffee in the Carlesberg Mansion, occupied by a selected scientist, like Niels Bohr. The scientist in the mansion who hosted us at dinner was Niels Bjerrum (1879-1958), the father of Jannik. From left to right are K. A. Jensen (1908-92), me, Asmussen (1903-84), Arthur Adamson, and Jannik Bjerrum.

---

[*]Niels Bjerrum was the scientist that occupied the Carlesberg mansion (**Fig. 2-6**). Carlesberg beer is transported globally and is said to be one of the very best beers. The last member of the Carlesburg family established the Carlesberg Foundation for the support of the Arts and Sciences in Denmark. The Carlesberg Foundation also selects the most outstanding Dane in the Arts or in Science to occupy the Carlesberg mansion. The Danes said Niels should have won the Nobel Prize. Earlier, the mansion was occupied by physicist Niels Bohr (1885-1962), known for his widely accepted theory of how electrons orbit the nucleus of an atom.

Some years later the Tuborg brewery, also said to make one of the very best beers to be sold worldwide, arranged to have its earnings contributed to the Carlesberg Foundation. Anyone drinking either beer will derive satisfaction from having a good glass of beer, but even more so by contributing to the Arts and Sciences in Denmark. While in Denmark, I did my very best to support the Foundation by drinking more than my share of their good beer!

Figure 2-7. Jannik Bjerrum's research group having lunch in 1954-55.

his students and assistants discuss it. Werner's coordination theory had the supreme distinction of being discussed only in the men's room by students of Jørgensen, out of his ear shot.

Our host Jannik had a very excellent research group working in different areas of coordination chemistry. These included Christian Klixbüll Jørgensen and Claus Schäffer on experiment and theory, Carl J. Ballhausen on quantum mechanical calculations, and Flemming Woldbye on asymmetric metal complexes. Other members of the department seemed to be more on their own (**Fig. 2-7**). This picture shows most of the group having their "brown-bag" open-face sandwich lunch together. Since we all came together each day at lunch and at afternoon tea, it was easy to get to know each other and to keep in touch with research in progress.

Likewise, Mary and I invited a few of the younger members of the research group to our house for dinner. They were happy to learn that in the US professors often have research students and assistants to their homes for dinner—not so at that time in Denmark. We all found it to be a worthwhile evening, with our talking about various subjects. However, one concern uppermost in their minds was that they would not become professors in Denmark until the present professors died or retired. This would be true in spite of their doing high quality research and publishing outstanding scientific papers. I tried to convince them that if they stayed the course, their time would come. It may have taken some longer than others, but they did eventually get appointments as professors.

During the year I collaborated with Jannik and Carl on a study of the absorption spectra of geometrical isomers of hexacoordinated complexes. In addition, Art and I investigated the H/D exchange of $[Co(NH_3)_5Cl]^{2+}$ (see base hydrolysis, pp. 88-91).

Although the lab was antiquated, it did have one piece of equipment not in my lab at NU. This was one of the very first Cary spectrophotometers that would mechanically take and record solution spectra. In my lab we were still doing this manually, one point at a time. Soon afterwards, however, our lab acquired two of the latest Cary models. The Cary in Jannik's lab had a tag on it with "courtesy of the Carlesberg Foundation" (see footnote, p.40).

Klixbüll Jørgensen was the person using their Cary. I am at a loss as to how to describe Klixbüll. He is an unreal character, perhaps even a genius. He speaks several languages, none of which are easy to understand. One thing about him that amazed me was how often he could recall a specific reference (journal, vol., page, year) of a paper in the literature that he had read. Each day when he arrived in the lab, he would prepare a water solution of some transition metal salt and use the Cary to measure the solution's ultraviolet-visible spectrum. He then would interpret the spectrum, making use of the "crystal field theory" introduced to chemists in 1951 by F. E. Ilse and H. Hartmann (*Z. Physik. Chem.*, 197, 239 (1951)), but known by physicists since 1930. Klixbüll's lab space was always a mess, he never labeled any of the containers of the solutions but he always seemed certain that he knew what each were. He was such a strange individual that I felt I had to believe him. Almost every Monday he would arrive in the lab with a paper he had written over the weekend. I was asked to read it to help with his English —which needed help. Not being a theoretical chemist, this was the first I heard of the crystal-field theory. It has to do with the electronic structure of the transition metals, and accounts for the visible color of their solutions on the basis of the d-d orbital transitions. The theories of bonding in metal complexes have now advanced from the Pauling valence bond structure to the modern molecular orbital theory.

Figure 2-8. Book by Fred Basolo and Ralph G. Pearson, *Mechanisms of Inorganic Reactions* (John Wiley & Sons Inc., New York, 1st Ed., 1958, 2nd Ed., 1967).

I did find time during the year in Copenhagen to write a couple of chapters on what was later to become the book by Ralph G. Pearson and myself, *Mechanisms of Inorganic Reactions* (John Wiley & Sons, Inc., New York, 1st Ed., 1958; 2nd Ed., 1967; **Fig. 2-8**). Ralph and I were fortunate in that the timing for the publication of our book was perfect. First, the crystal field theory for chemists was still relatively new, and there was a need for a book stressing its importance in explaining some of the properties of transition metal complexes. Second, inorganic chemists were beginning to investigate the kinetics and mechanisms of redox and ligand substitution reactions of metal complexes (see ligand substitution, pp. 84-85). A good book will have its maximum impact if published at the right time when it is most needed. That proved to be true of this book which got rave reviews. Daryle Busch, one of the reviewers and a good friend of ours, even said the book was destined to become the "Bible" of mechanisms of inorganic reactions—the term at times is still used by some chemists old enough to remember the review. The book was used globally, often referenced, and translated into German and Chinese. The first edition was easy for Ralph and me to write, since the literature on the subject was limited and we were familiar with it because of the nature of our research. In fact, some of the literature was from our own lab. The second edition nine years later was much more difficult, because the field had grown so rapidly that we were no longer as familiar with the entire literature. Needless to say, when asked to write a third edition, neither Ralph nor I were willing. The book is no longer in print, but at times I am told by some professors that they use their copy to prepare lectures for a course on the subject. It is always rewarding (not in a monetary sense) to write something that your peers and their students make good use of. One amusing comment about Ralph's and my long joint publications is that his physical chemistry friends, after a few years, would ask him "When are you going to give your student Basolo his degree, and stop taking advantage of him in your research group?" The same, in reverse, was being asked of me by inorganic chemists regarding Ralph. Later, they were all to learn the truth that our department had this unique feature to permit faculty to share students and do research together over long periods of time.

Now, back to Mary, MC, Freddie, Jr., and me at our home in Copenhagen. Later in the year, after having lived there for several months, we and the neighbors began to feel very close and comfortable together. They were very helpful to Mary and the children, and spoke only Danish to the children who were beginning to understand and speak a few words. The neighbors would tell us what we might want to visit or do that the children would enjoy. One obvious attraction for the children was Tivoli, which is an amusement park with all sorts of activities taking place. Everything is outdoors. There are few, if any, rides, but one can move from one outdoor stage to another, with something different taking place at each stage. One might have a humorous pantomime play, intended to entertain children; another a magician, again for children; high wire gymnastics; etc. Given a choice, the children always chose to go to Tivoli.

We also managed to take a few trips to near-by islands, taking a car ferry across. Since we often expected to be gone a few days, we had to tell the milkman, the butter and egg man, and the postman that we would be back on a particular day. Mary and I could not do this in Danish so one of the children would act as our interpreter.

One of our neighbors, who was very outgoing, said he wanted to take advantage of our being there. What he meant was that he wanted to practice speaking English to an English-speaking person. He had read much English, and had attended many English movies to understand spoken English. (This is the very reason the Danes do not dub movies in Danish, so that the Danes may learn to understand English.) Our friend said he felt sure he could speak English well enough to be understood, but had not had a chance to try it. Now he did with Mary and me, and he was delighted to know that he was correct, and that we could understand him. Every chance he got he would come to speak English with us. As all this shows, we were easily accepted by the Danes, and all of us loved it.

After one or so months, we bought a small English-built Ford, but the parts were assembled in Copenhagen. Visitors paid no tariff on the import of a car, but a Dane had to pay about three times the worth of the car. After driving the car all year, Professor K. A. Jensen said he would like to buy it because he would thereby avoid paying the high import tax. He asked me to go with him to get permission to do this. He told the officials that I needed the money to take my family back home to the US. This was granted and he wanted to pay me for what I paid when I bought it new. I insisted he pay less and we agreed on what seemed reasonable.

I drove each day to the university and parked nearby, just in front of a liquor store. This made it convenient to pick up a six pack of their excellent Carlesberg or Tuborg beer. I noticed that the natives often drank a very cold, hard liquor called "akvavit" which I did not like, mostly because of its poppy seed flavor. I decided I should learn to like it so I bought a bottle and had one drink every night. By the time the bottle was half empty I began to like it, and now I enjoy it very much.

Having a car, Mary and I decided to drive to Rome, stopping to visit other countries on the way. Neither of us had ever seen Europe, so we wanted to do as much as we could on the trip. This meant leaving the children behind, and that immediately caused a problem. We had been using a middle-aged lady who lived on our street to babysit at times. She and her husband, Mr. & Mrs. Pederson, were very fond of our two children. They had no children, so they enjoyed playing with MC and Freddie, and teaching them some Danish. They had a large dog that our children liked, and that was also friendly to them. When it came time to find a person to move into our house while we were to be

Figure 2-9. E. O. Fischer at the Technical University of Munich (TUM) who shared the Nobel Prize (1973) with Professor Geoffroy Wilkinson of Imperial College London "for their pioneering work, performed independently, on the chemistry of the organometallic so-called sandwich compounds."

Figure 2-10. Professor Walter Hieber (1895-1976) at the TUM, known as the "Father of Metal Carbonyl Chemistry."

gone, Mary asked the Pedersons if they could recommend someone to take care of our children for a month while we drove to Rome and back. They immediately said, "Oh no, you must let them come stay with us. It would be a big pleasure for us to have them and you just cannot say no, for if you do, we would be angry."

We felt very much at ease having our children stay with them. We telephoned every other day and were always told that they were no trouble. As it was summer, there was no school and they played every day with children their age. Later we were told that the father of one of the boys Freddie played with was Aage Bohr, the son of Niels Bohr, perhaps the most famous physicist other than Einstein. Of course, Niels was a Nobel Prize Laureate (1922), and Aage also received the Nobel Prize (1975) a few years after we had left Copenhagen. Unfortunately, none of this playing with Aage's son rubbed off on Freddie, but we love Freddie dearly as a high school science teacher in Bloomington, Illinois.

Mary and I made no hotel reservations before leaving on our trip because we did not know when nor where we would arrive. Neither did people we visited at universities know that we would arrive, so our appearance was always a surprise. After a long tiring day, our first stop was in Hamburg, Germany. All we wanted was a bit to eat and a good night's sleep. From our hotel room we could see a movie house directly across the street. We thought that the American movie would be in English with German subtitles, as it was done in Denmark. This was not the case in Germany, where the spoken English was changed to German. We were not able to understand German, but we did get our money's worth when we heard Geronimo speak German. From Hamburg we went to Amsterdam, where we stayed in a pension for the first time, and we liked it, mostly because one could have meals with others, enabling us to make some good lifelong friends.

I attended the scientific meeting of the International Conference on Coordination Chemistry (ICCC) for four days, while Mary took part in the sightseeing program arranged for spouses. I found this a professionally stimulating time for I met several inorganic chemists whose names and research I knew only because of reading their scientific publications. The same was true for them. My good fortune was to meet E. O. Fischer (**Fig. 2-9**) and we two became good professional and personal friends. It was very exhilarating to both of us to have a stimulating discussion eye-to-eye about research of interest to both of us. Likewise, I was very pleased to hear the plenary lecture of Walter Hieber (**Fig. 2-10**) and to have his helpful responses to my questions following his lecture (see organometallics, pp. 100-108).

After leaving Amsterdam, our next stop was Brussels, where we spent two days seeing most of the tourist sights. Little did I know that during the 1970s I would be going to Brussels twice a year, over seven years, to meet with a NATO research funding panel (see pp. 147-151). From Brussels, we went to Munster where I knew about the famous Professor Klemm (**Fig. 2-11**), later to become President of the International Union of Pure and Applied Chemistry (IUPAC). We went to the university where we located his Institute. When I asked if I could see him, I was told that he was on holiday and would be gone for a couple of weeks. However, his very fine young deputy, Professor Rudolf Hoppe, hearing I was a professor of chemistry in the US, was very helpful.

He showed me through the lab, explaining what was being done and why. Their solid state research was not familiar to me, because my research dealt with solution chemistry. Some years later, Klemm was my guest in our chemistry department (see pp. 192-193).

Our next stop was Paris where, on the first day, we behaved like American tourists and took a sight-seeing bus to see Paris in one day. The following day we spent with Professor Arthur Martell (**Fig. 2-12**) and his wife. This meeting had been prearranged, since we knew we were to be in Paris at the same time. Art and I are both coordination chemists, and we have been the best of friends since his early days at Clark University (now at Texas A&M). We had a leisurely day together, partaking of good French meals and wine. The wives compared notes on a sabbatical in Copenhagen vs. Zurich. It was decided that life for a foreigner is much better in Denmark than in Switzerland. The Danes are very friendly and helpful, whereas the Swiss are more formal and less likely to make friends. At the university level, Art and I found things to be similar and very satisfactory.

Mary and I then had a long, leisurely drive to Marseilles, with two overnight stops at some small villages on the way. We wanted to arrive in Cannes (near Nice on the French Riviera) to visit my mother's old friend, with whom she had exchanged letters over the years. Mother's friend, Maria Giorsetti, was caught completely by surprise to have Mary and me stop to see her and explain how her good friend in the US was my mother. She was a widow with three sons and a daughter. The daughter and one of the sons spoke English which was good, because neither Mary nor I knew French. The family operated their own small hotel and restaurant near a beach on the Mediterranean. They lived a short distance inland on the side of a hill with an excellent view of the sea. There they had a sizeable garden with jasmine plants, the flowers of which were harvested to make very expensive perfume. They wanted us to stay several days, but we told them that we could only stay the one day because we still had much traveling to do. That evening, when they were busy working in their restaurant, Mary and I had our first opportunity to sit on the beach of the French Riviera and watch the beautiful sunset.

Figure 2-11. Professor Wilhelm Klemm (1906-48) at the University of Munster where he had a large research group doing research on solid state inorganic chemistry. (At that time there was little such work being done in the US.) He made some interesting complexes with high oxidation state metals, such as $K_2[NiF_6]$ and $KCuO_2$.

Besides the sunset, the other thing that caught my attention was a game being played by some Frenchmen among the rocks along the outside of the beach. They were using metal balls about twice the size of a baseball, and were trying to throw it to go close to a smaller anchor ball. Their opponent would do the same, and points were scored by whose ball (or balls) were the closest. At some point they would even try to knock one of the other person's ball away. I did not understand how they could even expect to make any accurate shots among so many rocks. After watching for awhile, I was proven wrong and some of them were really able to make excellent shots. The French call this game *jeu de boules*. The Italians have a similar game called *boccia*, which they play on a very smooth surface or on a rough terrain. I have played the Italian *boccia* with some good players, and it is not easy.

Figure 2-12. Professor Arthur Martell of Texas A&M University, noted for his work on the stability constants of metal complexes in water solution and for compiling such data from laboratories worldwide.

Some weeks before starting on our one month drive in Europe to Rome, I had written to my sister asking her to send me the name and address of mother's best relative in Cuneo. Maria DeGovani and my mother had frequently written to one another so I was certain that my sister would locate the address. I then wrote to Maria, introducing myself as Catherina Basolo's son, and explained that my wife and I were coming to Italy and would like to visit them. Within a week I had her response, telling me how delighted she and our other cousins were and that they were looking forward to welcoming (benvinuto) us. Letters began to arrive each day for a week saying that they too were relatives, just as Maria, and so we must also visit them. The few days that we spent with relatives who could not do enough for us seemed like a perpetual Christmas day. The food and wine were outstanding and the stories they told us reminded me of what I had heard from my parents.

I had purposely taken the route via Cannes to visit mother's friend, but also to enter Italy from the French border. This put us immediately in the northern region of Italy known as Piedmont (Piemonte), where our next stop would be at the home of Maria DeGovani. Since we had a language problem as we traveled through Europe, I said to Mary, "Soon we will be in Italy, so relax, for I will take over our language trouble." At the border, the young custom person (**Fig. 2-13**) knew only enough English to say "Passport, please." I gave it to him, and asked him a question in my very poor Italian. He began to laugh, but then answered my question. I asked, "Why did you laugh, for my question was not funny?" He said because I figured you were Americans from your dress, but you are driving an English car with a Danish license, so I could not be sure of your nationality until I saw your passport. I did not expect to have you speak Italian, and certainly not the peasant dialect "piemontese" in this "piemonte" region."*

At last we arrived in the beautiful small town of Cuneo near the French border, which is the most wonderful village I have ever seen. My cousin Maria and her family were all joyously excited to see us. The excitement was not only because relatives had arrived to visit, but also because we were the first Americans to ever have been in their house. Mary knew no Italian, and they did not know English. Additionally, I found that my piemontese was not as good as I had expected it to be. It had been 20 years since I lived at home in Coello and spoke the language fluently each day.

The first few days I managed to communicate with the adults, but found it more difficult to do so with children who talked fast and expected an immediate answer. However, my piemontese improved day by day and soon I was able to think and speak Italian directly without having to think in English. At this point Mary and I began to really enjoy our visit. Since I could quickly serve as translator, Mary could get involved in the conversations. One thing I wanted to see was where my mother had lived. They drove us there, and the owners of the house very graciously invited us to have a look at the house, indoors and out. It was just as mother had described it to me. I was even shown a large smooth stone near a stream of clear spring water where my mother and grandmother washed the family clothes. This country region is called Tetti, and it is near the small village of Dronero. Such villages, no matter

Figure 2-13. The customs person looking at our passport as Mary and I enter Italy from France, 1955. This was in northern Italy, in the Piemonte region. I totally surprised the customs checker by speaking some piemontese, the dialect of that region.

---

*It was of interest to us that European countries have small areas of the country speaking the language of their area. However, this is different from our southern drawl, New York accent, etc., which allows us sometimes to detect where they are from. The words we use in the US are the same, although the accents may differ. In Europe, even some of the words often differ, so the Italians can immediately tell where a person is from. For example, the word for chair is "sedia," in the legal Tuscan language but it is "caderega" in Piemontese. This big difference dates back to when travel and communication was difficult and slow. However, now it has been a few generations later and the national language has been taught in all schools, and used on radio and TV. One wonders why these different languages are still used although they all know the same national language. I have asked some of my Italian friends about this and always get the same answer, "The people take pride in the region where they live and wish to continue this tradition."

matter how small, always have a Catholic church. It was in this Dronero church that my mother was baptized, and where mother waited to have me baptized. We entered the church so I could see the Baptism font. The priest saw us and came to ask if he could be of any help. We told him how we happened to be there, and he proceeded to show and tell us much more than we wanted to know.

Maria was an outstanding cook who prepared dishes that I liked and that reminded me of my mother's cooking. Her husband was a large, heavy man who had a large successful business selling groceries to the mom & pop retail stores. He was considered one of the wealthiest persons in Cuneo, yet he liked to tease me about "you rich Americans." One evening, after a good big dinner along with some good piemonte red wine—the red wines in this region are among the very best in Italy—he started on his rich Americans. I took my billfold, put it on the table, and asked him to do the same so we could trade even. He obviously would not do this, but after that, he stopped making remarks about rich Americans.

One other thing that they were curious about was the color problem in America and why we discriminated between white and black people. I had no good response to give them. It was clear that they had read about this over the years in their newspapers, because they mentioned the problem just currently happening in Arkansas. They could not understand how and why the governor of the state could have the national guard come to assist in having a segregated elementary school not permit black children to attend. They further did not understand how President Dwight Eisenhower could threaten to send the army to prevent this, should it happen. Fortunately, the governor did not call the National Guard, and black children were permitted to attend the school. However, they continued to ask "Why?" I told them, unfortunately, this all started years ago when blacks were slaves, without the right to vote. Now they do vote, do have integrated schools, do work with whites, etc. The problem of discrimination has come a long way and continues to improve, but it still has work to do to achieve the statement in the US constitution: "We decree that all men are created equal, so help you God." During the few days we were there, I heard them make disparaging remarks about Italians in the south of Italy (Sicily, Calabria, etc.). The problems between the northern and southern Italians are similar to our black-white problem. I called attention to this problem within Italy when they next asked about our situation, and it helped them to understand the predicament that the US is in.

Our next stop was in the smallest of towns (Borgiallo) where my father was from. The town is on top of a hill, providing a wonderful view of the surroundings below. It has a church and a few small shops. We did not have the address of any of my father's family, so they did not know of our arrival. We were given directions by the priest to my father's old house. We went there and got no response when we knocked on the door, except for a dog that started barking. There was then a call from the nearby vineyard asking, "Who is there?" Mary and I walked out to where they were tying branches of grape plants to prevent them from breaking by the weight of the bunch of grapes as they grew larger. I explained who I was, and that Mary and I came to see where my father had lived. My father's sister said, yes, this was the place and that my father had grown up in this house. She further said that the young man working with her was her son, and that they always had to do much work. During our brief conversation, they kept on tying the grape vines. It was clear that they were not interested in us, and did not offer to show us around. I asked if it would be OK if we had a look at the place on our own. It was OK so Mary and I walked over to have a closer look at the house from the outside, and the property surrounding it. Unlike my mother, my father had rarely talked to us siblings about his childhood. I gather that the family were not good Catholics, and the priest had little to say about them.

Our next visit to relatives was very sad. This was a visit to my father's brother, Tony, in Pinerolo. He and his family were our nearest neighbors for several years in the US. He and Aunt Rena had only one child, a son Aldo (see **Fig. 1-3**), who was one year older than me. Uncle Tony, a coal miner, had lost one eye by an accident while spraying his grape plants with a solution of copper sulfate. This meant that he could no longer work in the mine. As mentioned earlier, his father was a cattle merchant so Tony therefore decided to go to Italy to inherit his father's business. Aldo went to college and received a degree that would permit him to teach in an elementary school. However, in order to do this, he had to renounce his US citizenship and become an Italian citizen. This he did and, soon after, the Italians were involved with the Germans in WWII. Aldo was drafted and later captured and held prisoner. At the end of the war, he was released. However, he was very ill and, returning home on the train, he died before he arrived. When his parents saw me, they both burst into tears, and hugged me as if I were their son Aldo. They blamed themselves for his death, and kept saying it was their fault. If they had stayed in the US, then their son's death would not have happened. I told them that their son would have been drafted in the US, and thousands of Americans died in the war. He might have been one of the thousands, so it is not known what may have happened had they stayed in the US.

In addition to this sadness, it was clear that they were destitute, living in a small one-room apartment with hardly enough to eat. Mary and I insisted that we wanted to take them to dinner. They finally agreed to go, but they kept talking about their guilt for Aldo's death. They would say that since Aldo and I grew up and played together, he could have been just like me—a fine young man—had they just stayed in the US. Mary and I were pleased that this was only a short visit, for it brought tears to my eyes when I recalled how they were such a happy family when in Coello.

From there our last stop with family in Italy was in Genova, the home of Christopher Columbus. After our experiences at my father's home and with Uncle Tony, we were delighted to find a completely different situation at my cousin's house. My cousin, Enrico Marino, had my mother's maiden name. He was a very successful fruit merchant, and also owner of an artichoke plantation in Sardinia. He would awaken at 3:00 AM to supervise the unloading of large trucks with fresh fruit from southern Italy which would then be loaded into smaller trucks. These trucks would then go out each day to distribute the fruit to the small retail dealers. The early death of Enrico left the fruit business to his son Mino. He did not like that type of work, and started to build custom boats. As he and I walked along the beach, he would proudly point out to me, "That's one of my boats." We were only to be there two days, so the evening we arrived they planned what we were to see. Since Genova had been the home of Columbus, we expected them to start with something about Columbus. Instead, it was decided that we must see the graveyard. Mary and I looked at one another, and I asked whether this was their idea of a joke. They said that they were serious, because the cemetery, with its beautifully carved stones, was the best graveyard that one would ever see. We went the next day and had to agree that it was something worth seeing. The city had carved a large stone of a poor lady who for years, regardless of the weather, stood at the same corner in the city begging for money. A locally well-known carver made what is a spectacular stone likeness of the lady of which Genova is very proud.

Our next stop was in Milan where I wanted to meet Professor Lamberto Malatesta (**Fig. 2-14**). Malatesta translates into "bad head" which is certainly not applicable to Lamberto who was, and is, an outstanding chemist. There had been considerable destruction during WWII, and some of Lamberto's labs were under repair. We had a good visit talking in English/Italian about our research of

Figure 2-14. Professor Lamberto Malatesta at the University of Milan. He is one of the Italian chemists who played a major role in establishing outstanding inorganic chemistry in Italy. He was responsible for the syntheses of some important organometallic compounds of Ru and Ir.

mutual interest. I met his research group which included two young ladies who were doing outstanding work on the coordination complexes of rhodium (Rh) and iridium (Ir). One had prepared the complex $[Rh(PR_3)_2COCl]$, later to become known as the Vaska compound because Professor Lauri Vaska at Clarkson College of Technology, New York, discovered its oxidative addition and reductive elimination reactions. These reactions were to play a most important role in organometallic chemistry.

We then drove back using the coastal route with its beautiful views of the Mediterranean. In Pisa we stopped only long enough to see the Leaning Tower. Next was Florence, where we stayed two days with our gracious hosts, Maria and Luigi Sacconi (**Fig. 2-15**). We had never met either one, but Luigi and I had read each other's scientific papers. Some of his papers were published in English in the *Journal of the American Chemical Society*. I telephoned him as soon as we got situated in our hotel. I told him who was calling and where we were. I did this in English, thinking that since he published in English, he must know the language. He immediately said, "Let's please talk Italian, and I will be at your hotel in a half an hour." He arrived all excited that Mary and I were in Florence. He told us about the plans for that evening, and we gladly went along with his plans (later I was told that Luigi always expected to get his way with his requests).

Mary stayed at the hotel while Luigi and I went to his lab. He told me that he was just reading our latest paper on some nickel (Ni) complexes. When we got to the lab, he started by showing me the latest *JACS* on his desk, opened to our article (later I learned that this was most likely staged). In Italian, he told me all about his current research. Later the research resulted in his polemic with Professor Richard Holm, a formidable inorganic chemist now at Harvard University. With his poor English and my poor Italian, we managed to communicate. Mary, Maria, Luigi, and I became the best of friends and, on several occasions, were guests in each other's homes.

Our next and final stop was Rome. While there, I met Professor Vincenzo Caglioti who was the most "powerful" inorganic chemist in Italy (**Fig. 2-16**). We had a good visit, with me speaking my poor Italian because he knew no English. I told him I was on sabbatical at the University of Denmark with Professor Jannik Bjerrum as my host, during the academic year 1954-55. He invited me to take my next sabbatical in Rome with him as my host. I told him that I did not know if there would even be a second sabbatical but, if so, I would let him know. That was not enough for him, and he made me promise that my next stay in Europe would be in his Institute at the University of Rome. I learned later that Vincenzo was not accustomed to taking no for an answer.

Our return to Copenhagen from Rome was done as quickly as we could, with only hotel stops at night. After a month of travel, we could hardly wait to see our children. Both were very excited about seeing us, and rapidly tried to tell us everything that had happened while we were gone. They talked to us in English, but it was mixed with Danish. The Pedersons said they used only Danish with the children, and seemed to get along fine. They told us how happy they were with the children, and thanked us for letting the children stay with them. We were in Denmark for two more weeks before leaving again for a month, driving through Sweden and Norway to get to England.

We stopped only briefly in Sweden and Norway. I recall that although this was in late July, the snow covered mountain pass to Bergen was just being opened. This was done by clearing a road the width of a car, and at some distance from where it was possible to see an oncoming car, they opened a large enough area off the road allowing one car to get off the road so

Figure 2-15. Professor Luigi Sacconi (1911-97), at the University of Florence, his wife Maria, and I when we first met in Florence in 1955. Luigi's contributions to coordination chemistry were outstanding, and were particularly responsible for the worldwide attention given to Italian chemistry.

Figure 2-16. Professor Vincenzo Caglioti, (1902-99) at the University of Rome, is shown here greeting Antonio Segni, President of the Italian Republic. Caglioti was a very good inorganic chemist, surrounded by a fine research group. His greatest strength was his ability to bring the importance of chemistry, and other sciences, and the urgent need for its support and funding to the attention of the government. He later became director of the Consiglio Nazionale delle Ricerche (CNR), which corresponds to the directorship of the NSF.

that the other car could pass. We began to wonder if we would ever get to Bergen. Finally we did, and on the way we saw several herds of goats. Norway is known for its goat cheese that I would describe as having a tan/brown color with a slightly sweet taste. Our last night in Norway was spent in a hotel near one of its many gorges with a small waterfall near our room. Its pleasant noise made it easy to have a good night's sleep.

The following morning we took the ship, with our car on board.* We finally made it to Newcastle-on-Tyne in the UK. Our first stop was to be St. Andrews, which gave birth to golf. As we were driving up in that direction, Mary and I were talking about how we no longer would experience a language problem, since all the natives would speak English. On the way we needed some gasoline, and stopped at a station in Edinburgh

---

*The North Sea was very rough, as we were told it is much of the time. It was too dangerous to walk on deck, and dinner could not be served because the dishes would not stay on the tables. All of us became ill, including the ship personnel, due to sea sickness. Fortunately, the distance was not too long.

where I asked the attendant to fill it up. When he finished, I asked what I was to pay, and he told me. The problem was that neither Mary nor I could understand him. I put out my hand with some money and asked that he take what was needed. So much for knowing the language of our mother country! Later, when we told some of our English friends what happened, they said, "Oh yes, in that region they speak with a Yorkshire accent that even we do not understand."

In St. Andrews, we saw a bed and breakfast sign which was what we planned to use on our tour of the UK. The lady who came to the door said, "Yes, we have a room, but now we are all about to have our tea, so would you join us?" We did, and were ushered into a large living room filled mostly with people from nearby Glasgow on their holiday. When they saw that we were Americans, they, in a friendly fashion, began to ask questions about the US and also told us about their travels in the states. Finding out that we were from a suburb of Chicago, a few of them told us about a friend or relative who lived there. We did the best we could to respond but, once again, we had a problem understanding—this time, a Scottish accent.

I learned that one reason they came to St. Andrews was to see the most important British Open Golf Tournament (one of the majors, along with our Masters, Open, and PGA tournaments). I asked if I would be allowed to play on the Royal and Ancient golf course and, if so, where could I rent golf clubs and buy golf balls. They assured me that would be OK and told me that I could get clubs and balls at a small shop across from the pro shop. Early the next morning I did just what I was told, and I was the second person to tee off. The first person was teeing off on the second hole. I decided to rush a bit to catch up with him, and see if we could play together. From the 3rd hole through the 18th, we played together which was very helpful for me. The other golfer was on his vacation, spending it all playing golf each day. He had already played the previous two days, so he knew the positions of the hazards, and he could tell me where care was required. Even if one keeps his ball on the fairway, he may find a pot bunker filled with sand. Unlike visible sand bunkers, the pot bunkers are often not visible until you get near them. Since the person I was playing with knew the locations of these hazards, he would tell me and I would try to avoid them—not always with success! Both of us played about the same, poor to average, so we enjoyed our round together. When we were playing the back 9 holes, some of the professionals were arriving to practice for the British Open tournament the following week. Behind the 18th hole there was a temporary grandstand in preparation for the open match. Already there were some people there, awaiting the arrival of the practicing professionals. My friend and I had a good laugh, because they would be looking at us to see if they could recognize the two professionals. This was short-lived for they could see that we were just a couple of hackers. In any case, I was more than pleased that I had the opportunity to play golf on the course where the game was invented (**Fig. 2-17**).

Our next sightseeing stop was Glasgow in Scotland, where we spent two days and learned a bit more Scottish. Then we drove on to Leeds, where we met Jack Lewis and Ralph Wilkins, two young faculty members at the University of Leeds. They both were getting off to a good start with their research, and were destined to have outstanding

Figure 2-17. Me at the 18th hole of the Royal and Ancient Golf Course in Saint Andrews, where the game of golf started. Like most golfers, I looked forward to the day (1955) when I could play the course.

careers as inorganic chemists. One evening, the Wilkins invited Mary, me, and the Lewises to their house for dinner. We had a very fine meal with a lot of lively conversation, but then it came time to wash the dishes and clean up the kitchen. The time was about 9:00 PM and the pubs were open only until 10:00 PM. Us three men excused ourselves, while the ladies stayed home to wash the dishes. After a couple of pints of "warm" lager, we had a good conversation about the status of inorganic chemistry in the US and the UK. I recall their saying that it would not be easy for them to get a satisfactory faculty position in the UK because they had only attended the "red brick schools," not the more prestigious Oxford or Cambridge universities. I could appreciate this because in the US, at the time, it was likewise true that Ph.D.s in chemistry had a better opportunity to get a faculty position at one of the best research oriented universities if their degree was from Harvard, Caltech, or MIT than did an applicant of equal or better quality from a "lesser" school. Fortunately, this has changed for the better, both in the US and in the UK. For example, Jack Lewis became Professor of Chemistry at Cambridge, was elected a Fellow of the Royal Society (FRS), knighted as Sir Lewis, and now is Lord Lewis—how is that for one from a red brick school? Ralph Wilkins chose to come to the US where he became a distinguished professor and, in addition to his research, he wrote and has revised an excellent book on the kinetics and mechanisms of reactions of metal complexes.

We then went to Wales—another different language—and stayed at a charming bed and breakfast place. This was a house with only one room to rent. The elderly couple were very friendly and pleased to have us. They even wanted us to have dinner with them, and during dinner they asked about my parents and where I lived as a child. I told them that my father was a coal miner, and that we lived in a coal mining village of about 300 population. The two of them got all excited about this, because they too were living in a mining town and he was a coal miner. He then wanted to drive me out to see the mine. It seemed much like the mine my father worked in with a large tall smoke stack, and a large shaft to deliver the men to the depths of the mine and then return them to the top—much like an elevator, except there are no stops other than bottom and top. The next morning we went to Liverpool to catch a car ferry to Dublin. This was a necessary stop, because we had spent so much time in Italy where my parents came from. Mary is Irish, because her great, great grandparents had come to the US from Ireland. We took a sightseeing bus tour of the city of Dublin, then spent the night there, returning the next morning to England.

Our next stop, to visit a chemist, was in Nottingham. There lived Clifford Addison (**Fig. 2-18**), professor of inorganic chemistry at the University. He was, perhaps, the world's expert on the reactions and uses of dinitrogen tetroxide (N_2O_4), which is such a powerful oxidizing agent that serious explosions are commonplace. Again we had never met, but we felt we knew one another because we had read each other's scientific publications. He said he would call his wife to tell her of our surprise visit. She asked him to bring Mary to their house so they could get acquainted, and then he could return to his lab where he and I could talk chemistry. I had read a few of his papers on the synthesis and properties of volatile copper nitrate ($Cu(NO_3)_2$). This was quite an achievement, because we chemists only knew it to be a solid. He showed me his labs and introduced me to his graduate students. As we walked around the labs, I saw some tell-tale scars of what could have been an explosion due to the use of N_2O_4, but I did not comment on this. When we returned to Cliff's house, we found that his wife and Mary had gotten along so well that we were not only to stay for dinner, but also overnight in their guest room. They would not take no for an answer so we stayed and after a good dinner, we had pleasant conversation about children, US vs UK cultures, schools, etc. After breakfast they drove us around to see a few of the things they felt we would enjoy seeing. We then invited them to come as guests at our house when in the US. Cliff did come to give a seminar in our department, and did stay with us overnight.

Figure 2-18. Professor Clifford Addison (1913-94), at the University of Nottingham, who was noted for his career devoted almost entirely to the chemistry of N_2O_4, and for having prepared a volatile form of $Cu(NO_3)_2$.

On our way to Cambridge we stopped in Birmingham as tourists. The one thing we both remembered about this stop was attending our one-and-only cricket match. Mary and I both liked baseball, and she was a rabid Chicago Cubs fan. We went in alone, not knowing what to expect. They told us that an important, professional five-day match was in progress between England and India. There were very few spectators in the stadium, and, strange to us, spectators came and left whenever they wished at any point during the game. Additionally, the game was interrupted for a tea break for fans and players. One of the fans seated near us saw that Mary and I needed some help in trying to understand the game, and offered to help. He helped us to understand the gross aspects of the game, but the finer details were another matter. It was clear that the major matches were played over five days, explaining why fans came and went. The rest of the time they read the news to see how their team was doing. This is the only sport I know, except for golf, where one goes to see a game and leaves it before it is finished.*

From our delightful experience in Birmingham, we moved on to Cambridge University, one of the two major universities in the UK. There I met the famous inorganic chemist, Professor Emeleus. I was anxious to tell him how useful his book *Modern Aspects of Inorganic Chemistry* (H. J. Emeleus and J. S. Anderson) was for me when planning my lectures for a junior-senior level course on inorganic chemistry. He and his assistant were very British, but most gracious, immediately putting Mary and me at ease. They walked us around their beautiful campus, with all of its separate colleges. They told us about the history of Cambridge University, with its several giants of chemistry and physics over the many years. Emeleus had started his interesting research on the use of bromine trifluoride (BrF_3) as a solvent. He and his research group were discovering interesting acid-base reactions in this system. Years later, he visited us at NU and gave a seminar on this research, which had made substantial progress from

---

*It was of interest to me to test the difference in difficulty between our baseball (or softball) and cricket. In my early days at NU, I often pitched at our departmental picnics where the younger faculty played a team of graduate students and postdoctorates. We would, at times, even win a game, but as we faculty became older, while our opposition remained the same age, we no longer had a chance of winning, so we had to give this up. Dr. John Dickerson, one of my postdoctorates from England, and I would kid each other as to which was the more difficult sport, baseball or cricket. Of course, I contended it was baseball, because the game is played on a diamond occupying only a 90 degree area in which the batter must keep the ball, instead of the 360 degrees allowed in cricket. The baseball pitcher has several types of pitches to throw, and the bat is round, making it more difficult to have good ball contact than with the flat-faced cricket bat. For every point I made, he would counter with just an opposite interpretation. Our department had fall and spring picnics. John had arrived a few months before our fall outing, and he wanted me to pitch a softball to him so he could show me how easy it is to get a hit. After having thrown him several pitches, with some of our colleagues looking on, John was embarrassed because he was unable to hit the ball except for a few foul balls. John played county cricket in England, where he was the boller (pitcher). At our spring picnic, he arrived with cricket equipment and challenged me to be the batsman. I tried but failed miserably, because his pitches also differed as they bounced off the ground in various directions and at different speeds—I was not able to protect the wicket. We agreed that the difference in degree of difficulty is a standoff, and never again brought up the subject.

when he had briefly told me about it when we first met in Cambridge. Having "done" Cambridge, we now had to give equal time to the other major university, Oxford.

Here too we were not expected, but I was directed to the inorganic chemistry research labs. There I met Dr. Robert Williams who told me Professor Harry Irving was out of town, but that he could tell me about their on-going research. It was a nice day so Mary was outdoors, walking around the campus, waiting for my return. She had a long wait, because Bobby (as called by his friends) had a great deal to tell me about— not only about what research he was doing, but also what he planned to do. Finally I returned to find Mary sitting on a bench, and she asked "What took so long?" I told her that this young man was so enthusiastic about his research that he wanted to tell me not only all about what he knew, but also about what he didn't know. I told Mary this was an intelligent, ambitious young man who was surely destined to make his mark on chemistry. For example, there is an order of the stabilities of metal complexes of the divalent metals of the first row transition metals in the periodic table that increase regularly from left to right until it reaches copper (Cu), at which point it dramatically drops to zinc (Zn). In my opinion this should be called the "natural order of stabilities," but it is, and has been, called the "Irving-Williams Series." The order was even observed before by others or, at the very least, independent of Irving and Williams.

Mary and I were informed that the following day there would be the important Ascot horse races in Oxford, generally attended by the Queen. Mary and I went to it; the Queen was there; and the pageant and races were our first. It was a delightful afternoon for Mary and me to sit in the sun and watch the British elite, dressed in their finery, milling around and enjoying each other as well as the races.

Figure 2-19. Professor Sir Ronald Nyholm (1917-71), of the University College London with his wife and children. Ron was a very good coordination chemist, and an outstanding spokesman to laypersons on the importance of chemistry, and of science in general.

Figure 2-20. Professor Joseph Chatt (1914-94), at the University of Sussex, was one of our very best coordination chemists. He did some of the seminal work on the chemistry and bonding of platinum complexes. He became director of the government lab on nitrogen fixation, and he was awarded the Wolf Prize in 1981 for his contributions to inorganic chemistry.

Our next stop was London where we had our longest stay, saw the tourist sights, and attended a couple of plays. We stayed at a bed and breakfast place near the University College London. This was in "little Italy," and the owner and I would exchange some words in Italian. I was later to use this place rather often when I was on a NATO research panel (see pp. 147-151). Our host in London was Professor Sir Ronald Nyholm (**Fig. 2-19**), later to become one of my best friends. He had arrived from Australia a few years before to be appointed Professor of Inorganic Chemistry at the University College London. Not only was his chemistry outstanding, but he also was able to surround himself with the best young graduate students and postdoctorates who soon became the coordination chemists in England (see pg. 206).

Ron told me sometime later, when I was driving us around Piccadilly Circus, that he was a bit concerned about my having to drive on the left side of the road with a car for right-sided driving. There really was no reason for concern because, after the first few days, I found it easy to drive on the "wrong side" of the road. One thing that did frighten me was having to pass a large truck ("lorie") on a narrow two lane road. When behind a truck, this places the driver on the wrong side to see if there are on-coming cars. It meant that one had to depend on the person sitting next to the driver to see if it was OK to pass. Mary was that person and she managed to get us around with flying colors.

Our final stop in the UK was to visit Joseph Chatt (**Fig. 2-20**) at his Industrial Chemical Industries (ICI) laboratory at the Frythe where he was given complete freedom to do his own basic research. Joseph had to be considered one of the very best coordination chemists of his generation, and one who did some of the important seminal research on organometallic chemistry. For example, the olefin complex $K[Pt(C_2H_4)Cl_3]$, prepared initially by a Danish pharmacist in 1824, remained a mystery in terms of the platinum ethylene bonding. This bond type, making use of a classical σ-type component and a π component (see structure **15**), was proposed independently by Chatt and by Professor Michael Dewar, and is referred to as the Chatt-Dewar model of bonding. This

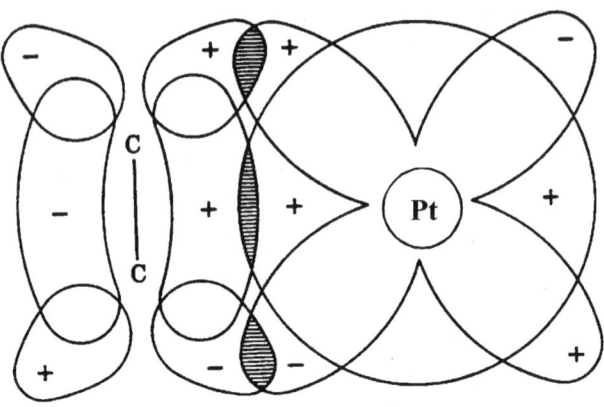

15

is only one of many seminal contributions made by Joseph, for which he received many Awards, including the Wolf Prize (1981) ranking in prestige below only the Nobel Prize which some of us believe he may appropriately have received. He and his wife Ethel took excellent care of us during our visit, and our two families became the best of friends, always exchanging long Christmas greetings.

Having completed our travel in the UK, and earlier in Europe, we returned to Copenhagen to prepare for the return to our house in Evanston. This involved packing many of the items purchased during the year, along with our original belongings. Because of the high dollar value relative to the Danish krone and because our rental of the house in Copenhagen cost half of what we received from the rent on our house in Evanston, we had saved enough money to buy some of the more expensive items we liked and continue to use today. These include Danish furniture, a 12 place setting of Royal Danish china, and a 12 place setting of Jorge Jensen silverware (stainless steel as the silver was more than we could afford). These things were all packaged and shipped to Chicago, without damage, where I got them through customs, put them in a U-haul van, and safely took them home. Mary and I enthusiastically agreed that this sabbatical had been a most worthwhile and rewarding life-time experience.

BACK HOME IN EVANSTON

We returned to our home in Evanston and, about a month after our return, we had a Danish guest for dinner who was able to easily speak with the children in Danish. Although he was able to talk with them in his native language, they had already forgotten much of what they had learned. We were told that this is the usual case with children; they learn a language quickly but also forget it quickly.

Seven years after the birth of Freddie, we were presented (1957) with a surprise in the form of the blessed event of the birth of our second daughter Margaret (Peg).

Because of the age difference between Peg and the other children, we felt she should have a sibling around her own age. We were hoping that the fourth child would be a boy so that we would then have two girls and two boys. In fact, as Mary and I were leaving the house for the hospital, MC and Freddie said, "Mom, if it's not a boy, you can't come home." Mary did come home, but with a girl, Elizabeth (Liz). We all loved Liz as our little baby who now, at age 40, has a beautiful family of her own: a husband and three children. Mary and I were blessed with four caring, delightful children and eleven grandchildren, all of whom we love very much and have always been very proud of.

Mary liked being outdoors during good weather. This way she got to know all the wives of our neighbors. I knew them just well enough to say hello. One of Mary's best friends was a Rickett of the Rickett Restaurants in Chicago, and a devout Catholic with seven children. She told Mary that they needed more room, and were looking for a larger house to buy. Mary and her friend struck a deal where they delayed putting their house on the market until we sold our house. With luck, this worked as planned, and we moved in their larger two story house just directly down the alley on Colfax street. The move was made easy because our friendly graduate students insisted they wanted to do it. They moved everything except the very heavy items. Besides giving us more room, the backyard was just in front of an excellent city park. It had sand boxes, swings, and such items for small children. It also had three tennis courts, and an 18-hole golf course. Our children made very good use of this park. When they became young teens, I taught them how to play tennis. As they became older, they became so good that I was no longer able to give them a good game. Margaret (Peg) did well enough to make the varsity team at UI. I was able to teach Freddie, Jr. how to play golf. I recall that when he finally beat me by two strokes, he ran to a pay phone to tell his mother. After that we played about even. Fred made the high school varsity team.

SABBATICAL IN ROME

In 1955, when Mary and I drove through Europe, one of our stops was in Rome. While there, I met Professor Vincenzo Caglioti who made me promise that my next stay in Europe would be in his Institute at the University of Rome. And so, in 1961-62, my family and I had a delightful year in Rome.

Our voyage to and from Italy was on the *Andre Doria*, the ship that later sank offshore near New York after a collision with another vessel. The six days on the ship were much the same as our travel to Denmark. I still had to take Dramamine, but the rest of the family enjoyed themselves, taking part in all the activities provided for passengers. The food, the wine, and the ship were beautiful. However, I was not enjoying it much and could not arrive soon enough. It was a shame to have it sink a few years later.

As we approached Naples, Mary and I planned our route from Naples to Rome, but we knew it would be much more difficult to find our home in Rome. All we had was a street address, which would not be very helpful since we did not know our way around Rome. Much to our surprise and profound pleasure, as we embarked the ship, we were met by Professor Caglioti and his wife. They had driven down from Rome to meet us because they realized that we would have a difficult time finding our quarters. We followed them, with some difficulty, not yet knowing how to drive in Italy. Finally we arrived at our "villa," a large two story house with a large yard. Our quarters were upstairs with the widow-owner occupying the bottom of the house. As in Copenhagen, we were fortunate that the previous tenants had recently left for a year, and their house had all of the things necessary such as dishes, bedding, etc.

Our second day there, Liz, our 2 year old, broke a small light bulb in her mouth. We feared that some of the pieces of glass got stuck in her throat, so we wanted to take her to a doctor. I contacted Caglioti and asked if he could recommend a doctor. He immediately said, "Bring her to me and we can take her to my doctor, because there are many bad doctors in Italy." This was done and we were relieved to learn that she was OK. All of these happenings the first two days told me that with Vincenzo as our host, we were in good hands. This was proven correct during our most pleasant year there.

This sabbatical year differed from the year in Copenhagen when we had only two children, Mary Catherine (6) and Freddie (4). Now they were ages 13 and 11 and, in addition, we had Margaret (4) and Elizabeth (2) (**Fig. 2-21**). Since there would, at times, be six of us traveling by car, we decided to take our small Ford Falcon station wagon. Although this was considered small in the US, it was considered large in Italy because most Italians drove little Fiats, "cinquecento-500." I learned to drive as aggressively as the Italians, and I had no serious accidents during the year. However, there was one fender-bender when a lady drove into the back of my car when I had stopped for a red light. She immediately got out of her car and started shouting at me that I was at fault. On such occasions, I always pretended that I knew no Italian. Two Italians, who were sitting at the front of a "bar" having coffee, came to my rescue. They shouted at the lady, telling her that they had seen what had happened, that it was her fault, and to stop making such a scene toward a tourist. She was driving a small Fiat 500 and had some damage to the front of her car, but nothing happened to my larger Ford Falcon.

Gasoline was expensive in Europe, but it is sold much cheaper to tourists for a period of six months. When I had been there six months, I mentioned this to one of my friends in the lab and he said, "Come with me to the vehicle office and pretend you only know English." We went, and my friend asked them to extend my tourist card for a second six months, because I was an American guest of the University. This was done so I had the full year with cheaper gas.*

---

*On one of my many visits to Italy, I was talking to a taxi driver. My Italian is good enough so that I can manage in ordinary conversation. He asked, "What are you doing in Italy?" I told him that I was a chemist attending a scientific conference. He said, "If you people are good chemists, you should be able to convert wine into gasoline—for gasoline is more expensive than is local wine." As I write this (October 2000), this is of particular concern with the high prices of gasoline globally.

Figure 2-21. Mary, me, and the children during our sabbatical year (1961-62) in Rome.

Another immediate concern after our arrival was to enroll MC and Freddie in an English school so that they would not lose a year of school. This had to be done immediately, because we arrived in September and school had already started. Again, I asked our host about this and he put us in touch with a staff member at the US Embassy. The staff person told us, "Yes, there is a very good English school used by American and English families in their Embassies and by members of NATO." He asked where we lived and I told him, "Montesacro, near the church's square." He said, "That is excellent, because the school bus picks students up at the square." Mary, the children, and I went to the school and made the arrangements for them to attend. This went smoothly and, at the same time, we got the books they would need. Mary, the children, and I also went to the square to meet the bus and tell the driver that MC and Freddie would be there each day to take the bus to school. That being done, we wanted to make arrangements for Peg. She, at age 4, could not attend school. Mary managed to have her go to a small children's care group under the attention of a Catholic nun. The group was cared for in the morning until noon. It was a short walk for Mary to take Peg to the church in the morning and to get her at noon. This meant that Mary only had Liz to look after in the mornings. Much of that time Mary spent shopping with Liz.

We lived in a suburb of Rome called Montesacro, in the top floor of a house on a hill. From our balcony, we had a wonderful view of the surroundings. All we could see were Italians who spoke little or no English, since there were no tourists in this area. I was able to manage with my Italian, but my family found it difficult and frustrating to communicate. Slowly they began to feel at ease with the problem, largely because Italians are very tolerant with foreigners and do their best to understand them. Mary could make use of a supermarket, but she preferred to buy groceries and other items in the small mom-pop shops. After we had been there for a couple of months, Mary became well known by the shopkeepers. In fact, when they saw Mary coming toward their shop, they would say "Guarda, arriva l'Americana" (Look, the American arrives).

When we arrived in September, it was the time to harvest grapes in the area. These are white grapes known as frascati. They make good wine, but they are also large, sweet, and very good to eat. Our family liked them and for some two or three weeks during the season, we ate more than our fill. The frascati wine is a good table wine, and I managed to drink more than my fair share during the year. The wine is generally available only in this area of Italy, because it is not Italy's highest quality white wine. A wine shop near where we lived had the wine in large wooden barrels and I would go there for my frascati wine. At times, I would have Freddie go to get me a bottle filled with wine. They would sell him the wine, even though he was only 11 years old.*

There is much to see and do in Rome, so we did what was possible for us. The first and most important thing for a Catholic is to visit St. Peters. We had shown the two older children pictures of St. Peters, but they were not prepared for it being so huge. One thing they did not see on that first visit was the Pope. We took them back and they saw the Pope giving Mass from his window and, later, outdoors. The children were pleased because they saw him in his pontifical robes, and he looked as they expected from having seen him on television.

One of the first things we got for the children was a TV because they enjoyed watching it at home and because we thought that listening to the Italian might help them to learn the language. Some American TV programs were often shown by the Italian TV. I recall MC and Freddie having me interpret one of the Perry Mason shows for them. I noticed that they were able to anticipate what was about to happen, and they voluntarily said they remembered having seen it when home in the U.S. One other thing about Italian TV at the time is that their programs were not interrupted with commercials. Instead, they had their commercials between programs, along with some entertainment for children. This was called "carosello" (carousel). It always announced itself with the same music and then showed some animated children's attraction using Mickey Mouse or Popeye. When our children heard the sound of that music, they would drop what they were doing and run to the TV set. The other thing that interested the children were the catacombs. Freddie, in particular, got me to take him there several times. In fact, during the year when we were visited by some Professors from the US, Freddie was even able to act as a guide.

Mary hired an Italian lady to come a few hours each day to help with the children and with taking care of the house, but also so Mary could practice her Italian. During the year, I was invited to spend a week at the Weizmann Institute in Israel to give a series of lectures, but with ample time for sightseeing. Mary had her Italian helper move in for the week to take care of the house and the children. Our friends at the Institute made

---

*This table wine cost about the same as did their bottled water, so I drank little water during our year in Rome.

certain that on our free day we were taken to see all of the important sights in Israel. In fact, near the end of our stay, having seen all the biblical places, our driver said he had a good Druse friend whom we could visit. Our hosts agreed that this might be of interest. We stopped at the driver's friend's house and they greeted one another with great glee. We were introduced to his friend who asked us to have tea. He called one of his three wives to prepare and bring us the tea. She was not introduced to us, nor did she stay to have tea with us. Later we were told that the Druse were allowed to have more than one wife, and our driver said that the husbands selected one each night with whom he would like to sleep.

There are two things that I recall most about this week in Israel. One was the visit to the Knesset, which one often hears or reads about and which corresponds to our congressional quarters. The other was being in my hotel after sunset on Sabbath, when the Jewish people do not eat nor travel. The only few cars on the streets are police or some emergency vehicles. I sat on the balcony of my hotel, overlooking the city of Jerusalem which had the dreary feeling of being dead. In the dark one could hear sounds never heard in such a large city—at some distance of children playing and dogs barking.*

When we returned from Israel, we asked the children what happened while we were gone. They said that nothing had happened. However, Freddie then asked, "Dad, why do they let Communists come to our square?" We wondered what he meant, and he said that they were bad guys. To our knowledge, he did not know anything about Communism, much less that they were bad. The only thing we could think of was that, even at age 11, children listen to grown-ups talk and come away with some of what is being discussed. The reason Freddie said that there were Communists with red flags making noise on the square was because there was an election for the Mayor of Rome and the Communist party, which is strong in Italy, had been electioneering for their candidate.

When alone in the house, Mary would not answer the telephone, because she would have a language problem unless the caller spoke some English. However, after several months with her Italian lady house helper, Mary began to make an effort to understand and say some words in Italian. One evening as I returned from the university and closed the door at the foot of the stairs, I heard a conversation going on upstairs. I quietly went up to hear better, without being seen. What I heard was that Mary and her friend were able to carry on a meaningful conversation in Italian. This was a turning point for Mary, as she no longer was frightened by the language problem.

In addition to what has just been described about our stay in Rome, I did have time for chemistry. Caglioti decided to arrange a program where every other week he

---

*I visited Israel on one other occasion. This was as a guest of the Wolf Foundation. I had been a member of the jury, along with two Nobel Prize laureates, who selected Professor Henry Eyring to be the 1980 Wolf Prize awardee. This prize in chemistry is viewed as being second only to the Nobel Prize. The founder of the Foundation was Dr. Ricardo Subirana Lobo Wolf who dedicated the latter time of his life to philanthropic work. We all met Wolf, who was then age 93. He died a year later, in 1981.

would invite a coordination chemist to his institute for a lecture on their research. Since I would be there for a year and since I am a coordination chemist, Caglioti felt this was an opportune time to give the people in his research group an exposure to this area of chemistry. He asked me to prepare a list of outstanding coordination chemists in Europe. The list I gave him was Professors Jannik Bjerrum, Joseph Chatt, Ron Nyholm, Gerhart Schwarzenbach, Luigi Sacconi, Lamberto Malatesta, Luigi Venanzi, and Klixbull Jørgensen. He invited each of these chemists, and their lectures gave Caglioti's research group the exposure to coordination chemistry he wished it would. Since he spoke no English, I was always asked to help host these guest speakers. This I was pleased to do, because I knew them all very well.

Some other chemists who came during the year I had never met and was pleased to meet. I recall particularly three of these visitors. One was Professor Klemm who, at the time, was the President of IUPAC. In this capacity he was in Italy to discuss the status of chemistry in Italy. Another was Professor Henry Eyring (1901-1981) of transition state fame. As always, his lecture was exciting, enjoyable, and also delivered an outstanding presentation of his research. Before the visitors' talks, Caglioti had them come to his office for coffee. Henry was a Morman, so he told us he could not have coffee. Likewise, he could not have alcoholic drinks. We were about to stop trying to think of something to suggest when it was said that orange juice may be satisfactory. Henry agreed that would be fine. A member of Caglioti's staff went to get the drinks and returned with the juice of a sanguini orange. This orange is red inside and is considered by Italians to be their best orange, so they take advantage of eating it when it is in season. Henry would not drink it, thinking it must be some kind of wine. After my effort to assure him that it really was orange juice, he finally drank it.

The third person was none other than Professor Robert Burns Woodward (1917-1979) of Harvard University. He was considered to be the best synthetic organic chemist in the world and obtained the Nobel Prize in 1965 "for his outstanding achievements in the art of organic synthesis." He was in Rome to accept his invitation to membership into the Vatican Academy of Arts and Sciences. This Academy is said to be the most prestigious Academy in the world. I had never heard Woodward give a lecture, but I had been told that he gave his lectures without the use of slides and talked for much longer than one hour. His lecture was long, as he wrote the chemical equations and structures of compounds on the blackboard. I did not mind his long talk because I liked hearing of his chemistry and marveled at his presentation —wishing that I could do as well. His talk had to do with the total synthesis of vitamin B12. Interestingly, several years later I was invited to Harvard for a couple of days to give three lectures on our research. Woodward attended only my talk on synthetic oxygen carriers, the one we had mostly studied were complexes of cobalt. After my lecture he came up to me and invited me to come to his office at 9:00 AM the next day. I did, and he said they were having difficulty putting cobalt, an important atom in vitamin B12, into their molecule. I said perhaps this was because the ligand substitution reactions of Co(II) complexes are labile, whereas those of Co(III) are inert. Since Co(II) compounds are readily oxidized by air to Co(III), I suggested that his group might try doing the reaction of Co(II) in an air-free environment. I do not know that this was ever tried, but Woodward did finally achieve the total synthesis of vitamin B12. One other thing about Woodward that was well known was the extent to which he liked the color blue. In fact, I did notice that every item in his office was blue, and it was referred to as the blue room. I was also told by some of his colleagues that I was fortunate to have been invited to the blue room.

When it was known by my research group that I would spend the year in Rome, one of my students asked if he could come and complete his dissertation research in Rome. Since he had completed all other requirements, I asked Caglioti if it would be possible to have Keith Stephen come to do his Ph.D. dissertation lab work there. I was told he was more than welcome. Keith and his wife Susan came, and they had a good year in Rome. Keith worked with a visiting scholar, and we published two papers on the paper chromatography of metal complexes. Professor G. Illuminate and I collaborated on research done by his student on the structure and reactivity of octahedral complexes, and we published a joint paper.

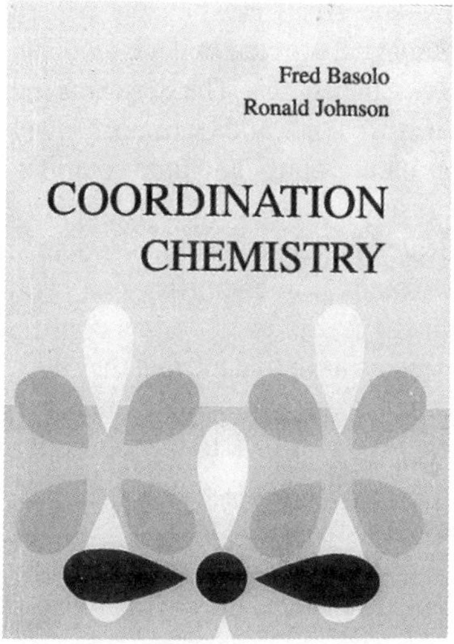

Figure 2-22. Book by Fred Basolo and Ronald C. Johnson, *Coordination Chemistry* (Benjamin Cummings Publishing Co., Menlo Park, CA, 1964). It was translated into seven different languages.

Some of my time during the year was spent in the library browsing through scientific journals. One of the articles I read gave me an idea for the syntheses of linkage isomers of the ligand NCS (see pp. 91-94).

Furthermore, my interest in teaching beginning students left me with the feeling that textbooks on general chemistry should be doing a much better job in the chapter on coordination chemistry. I decided to write a small paper-back book (**Fig. 2-22**) on this to supplement the existing textbook. The book would discuss crystal field theory, reaction mechanisms, stability, structures, and colors of metal complexes at the beginning student level. I wrote two chapters in Rome, and outlined what needed to be done in the remaining part of the book. When I returned home there was so much that required my attention that it seemed I would never be able to finish the book. I asked my former Ph.D. student, Ronald Johnson, then on the faculty at Emory University, if he would be able to finish it as co-author. Ron was interested in doing it and did an excellent job, even improving the two chapters I had written. The book, published in 1964, was entitled *Coordination Chemistry* (Fred Basolo and Ronald C. Johnson, The Benjamin/Cummings Publishing Co., Menlo Park, CA, 1964).

The small inexpensive paperback book was quickly and widely adopted for beginning freshmen college students, and even for high school students in advanced placement courses. The book was translated into Chinese, Japanese, Malaysian, Italian, French, Spanish, and German. On my worldwide travels over the years, I have had some prominent young inorganic chemists tell me that they had chosen inorganic chemistry because of our book *Coordination Chemistry*. In some countries, a student must indicate his field of interest—chemistry, physics, etc.—when he enters the university. For example, in chemistry one must then decide during the first year the branch of chemistry—inorganic, organic, physical, or analytical. Some of the now well-known professors have told me that they decided on inorganic chemistry after reading our book, which was easy for them to understand and made them believe it to be an exciting area of chemistry. It is a tremendous satisfaction for me to be told how our book had been so well received. It came at a time when there was a need for such a book, but now the subject is covered in general chemistry textbooks.

We enjoyed our year in Rome, but neither Mary nor the children learned much Italian. My Italian improved, going from piemontese closer to the real Tuscan Italian.

* * *

Our family then went through the usual things that families do, with many of these being delightful and interesting events to us. However, I will not burden you with so many details as above. We continued to visit Mary's mother each summer until she passed away. Once we went to Florida, and all of the family acquired a bad sun burn. Another time, we decided to drive and camp all the way to California, purchased all of the needed camping material, and stopped to see the Grand Canyon, Yellowstone National Park, etc. When we returned home exhausted, we gave all of our camping gear to a friend and asked that it not be returned. All of that camping convinced us that we would much rather stop overnight at a motel with a swimming pool.

Earlier the point was made that as the children grew up, I taught Freddie how to play golf and he soon was able to beat me. I also taught all of the children how to play tennis, but again, in a short time they became too good for me.

3

FACULTY POSITION AT NORTHWESTERN UNIVERSITY (NU)

SEPTEMBER 1946

At the end of World War II, many returning GIs had the cost of their education to a college or university of their choice paid by the government. Enrollment at these schools experienced a sudden, massive increase. As a result, schools quickly had to add new faculty, because of the large influx of students. Therefore, qualified persons were in demand, which meant that one could just about select the school where they wanted to teach and do research. As soon as the War had finished, I was prepared to choose my school, since I had long known that I really wanted a teaching position.

My choice of NU was strongly influenced by my desire to be in the Midwest, near family and friends. Another reason that I selected NU, rather than some other school in the Midwest, was because of its location. Having spent most of my life in a small village, I felt that it would be exciting to live near Chicago and to become an avid fan of the Cubs and the Bears.

In order not to create the wrong impression, there is a much more important reason why, after a visit, I was *sure* that NU was where I should launch my professional career. The reason was that it reminded me of my days as a graduate student at UI because it had only a half dozen students doing inorganic chemistry, while the remaining two hundred or so were mostly in organic chemistry. This was not because the inorganic faculty members at the UI were not as qualified as the organic faculty, it was only that inorganic chemistry had not yet advanced to its present status. And not only did I begin here as an instructor in 1946, but I remain here now, in spite of other attractive offers,

Figure 3-1. Charles D. Hurd (1897-1998), a well-known organic chemist, who was the senior member of the chemistry faculty when I arrived—and for long thereafter, for he lived to age 101. He is known for his research and for his book on pyrolysis of organic compounds.

in the year 2001, as the Charles E. and Emma H. Morrison Emeritus Professor of Chemistry.

The mystique of organic chemistry at NU was primarily because Frank C. Whitmore (1887-1947), of carbonium ion fame, had been a former faculty member. Frank was on the faculty of NU from 1920 through 1929. He then left to join the faculty at Pennsylvania State University, and was elected a member of the National Academy of Sciences in 1946. The department also had another well known organic chemist, Charles D. Hurd, who remained at NU throughout his professional career, from 1924 until his death at age 101 (**Fig. 3-1**).

At the time I was being interviewed for a faculty position, the chemistry department at NU was best known for its unique research on heterogeneous catalysis—unique, because it was the only chemistry department in the US doing this type of research. This work was largely being done by research chemists in the petroleum industry. Its importance was amply demonstrated by the improved production of gasoline and many other oil-based products.

SOME NORTHWESTERN HISTORY

It is of interest that, from its founding in 1851 to 1922, the only Ph.D. in chemistry was awarded by NU in 1896. As an inorganic chemist, I was pleased to learn that the research for this degree had to do with the hydrolysis of ferric chloride.

The catalysis laboratory at NU is an end result of the defection of V. N. Ipatieff (**Fig. 3-2**) from the USSR in 1930. Upon his arrival in the United States, Ipatieff was hired as a research chemist by the Universal Oil Company. Since he was one of the leaders in petroleum chemistry, he was given complete freedom to make his choice of research. After

Figure 3-2. V. N. Ipatieff (1867-1952) who did much of the seminal work on heterogeneous catalysis and, to his left, his assistant, Herman Pines (1902-96). Herman later became the Ipatieff Professor.

Faculty Position at Northwestern University

a few years, he decided that he would like to be an adjunct professor at an adjacent university. Northwestern University, being nearby, arranged to have him occupy a laboratory two days a week. He did not speak English, so he had his assistant Herman Pines come work with him at NU. Herman was able to speak several languages, albeit poorly but, fortunately, one of them was Russian. He became our first Ipatieff Professor, to be followed, in chronological order, by Robert Burwell, Wolfgang Sachtler, and Tobin Marks.

In 1932, Ipatieff and Pines showed that paraffin *iso*-butane reacts with olefins in the presence of strong acid. This was the basis of the alkylation process which was patented in 1938 and industrially developed soon after. Its most spectacular application is the synthesis of *iso*-octane from *n*-butene and *iso*-butane. *Iso*-octane improved the quality of gasoline and airplane fuel. It also played a decisive role in the victory of the Royal Air Force during the Battle of Britain in 1941.This was the beginning of catalysis research at NU, which is now widely known for its Ipatieff Laboratory. The Ipatieff Laboratory continues to do some of the best catalysis research in universities. Since catalysis has become a "hot" area of research, the NU lab is now joined by several other university labs.

Figure 3-3. Walter Murphy (1873-1942) was the philanthropist who gave NU the money (before WWII) required to build and maintain the first stage of our Tech building. He wanted to remain anonymous, so it was called the Technological Institute. The building is now about three times its original size, and it is called the McCormick Building.

I have been told that the chemistry department was originally housed in one half of old Fayerweather Hall, which has long since been torn down (physics occupied the other half). As the chemistry department increased in size and importance, additional space was provided by moving it into parts of Fisk Hall, Old College, University Hall, and a one room tin shack, which was used for glass blowing.

Fortunately, the housing problem of the chemistry department was past history by the time I arrived in 1946. The problem was solved by a donation of roughly 30 million dollars (worth much more today) to NU, which was given anonymously for the construction of an engineering-science building. It is said that the anonymous donor first went to the University of Chicago to make his donation, but had not been taken seriously. His treatment at NU must have been much better because this was where his gift was made. The anonymous donor is now known to have been Walter Murphy (**Fig. 3-3**), who made his millions by patenting railroad box cars which had been corrugated to give them added strength.

The building was named the Technological Institute and is still referred to as "Tech," although some years ago its name was changed to the McCormick Building. Tech was completed at about the same time as the end of World War II. Many universities were constructing chemistry laboratories and buildings during this period. Since Tech was the very newest technological building, we had several visitors from other universities, who came to see our facilities. They asked many questions about the latest construction and performance of our chemistry laboratories.

Between 1945 and 1947, six of us (Raymond Mariella, Norten Melchior, Robert Letsinger, Allen Hussey, and Ralph Pearson) arrived to begin our careers as instructors on the chemistry faculty at NU. Four of us survived and obtained tenure. My starting salary was $3,000 per year, whereas my Rohm and Haas salary had been $5,000 per year. This indicates how much I wanted an academic position. However, at age 26, I had yet to marry and had no family responsibilities. Even $3,000 for a nine month salary seemed adequate.

TEACHING

One reason that I was hired by the NU chemistry department is because it needed someone on its faculty to teach inorganic chemistry. Professor Harold Walton, one of the two inorganic chemists in the department, had left to take a position at the University of Colorado. His emphasis had primarily been on teaching and, soon after leaving, he wrote a useful book entitled *Inorganic Preparations*. His hobby, mountain climbing, was probably a motivating reason for his move to Colorado.

The other inorganic chemist on the NU faculty was Professor Pierce Selwood (1905-86), who had obtained his Ph.D. at the UI with Professor B S Hopkins, an inorganic chemist doing research on the lanthanide elements. Selwood was a graduate at the time Hopkins announced, incorrectly, the discovery of element 61, which he named *ilinium*. Ilinium was given the symbol Il (see Chapter 2, pg. 13). Pierce decided to use his training as a graduate student to do his own research, different from that which he had done for his Ph.D. dissertation. Even today, this approach to start one's career is advisable for, otherwise, the person will be viewed as having few or no research ideas of his own, and only capable of following the work which he had done with his mentor. As mentioned earlier, at that time the strength of the NU chemistry department involved research in the area of heterogeneous catalysis. Since very little was known about solid catalysts, Selwood decided to focus his attention on the nature of the solid surface of the catalyst. He was certainly a pioneer in this area of research, which now is referred to as chemical surface science, and is a very active area of research. Pierce primarily used magneto chemistry to examine these solids. This work prompted him to write a definitive book in 1943 entitled *Magneto Chemistry*. Two editions followed, and the book was widely used by inorganic chemists. However, he was never regarded as an inorganic chemist, consistent with other inorganic chemists in his age group. The net result was that there were very few to no inorganic chemistry graduate students when I arrived at NU.

Faculty Position at Northwestern University

As a beginning instructor, my position was primarily as a teacher to replace Walton, since graduate students were not selecting inorganic chemistry for their Ph.D. research. I was not unhappy with this assignment because as an undergraduate at SIN, I had taken several education courses, and had been an assistant high school chemistry teacher (see section on SIN in Carbondale, Chapter 2). Therefore, I knew I enjoyed teaching and looked forward to it, but I also wanted to have time to do research.

My first year on the faculty was largely spent teaching tutorial classes of about 25 freshmen. [At that time this was done by junior faculty. However, for many years now, it has been the responsibility of graduate student teaching assistants.] All of the lectures in this general chemistry course were given by Professor Robert K. Summerbell (1904-62). He devoted full time to his lectures and lecture demonstrations, and the students liked him and his course. He also expected the teaching of tutorial classes to be held to similar high standards. Although there were no student evaluations of courses and faculty at the time, student comments were very positive about both the lectures and the tutorial classes. Unfortunately, Summerbell had a heart attack and died an untimely death while attending a University Presidential reception (1963). At the reception, University President Roscoe Miller, a medical doctor, did all he could to keep Bob alive, but failed.

After Summerbell's death, his successor to teach general chemistry was L. Carroll King (1914-99). He was an organic chemist and an assistant professor. His research was going well, and it was supported by a grant from the Cancer Institute of the National Institutes of Health (NIH). Some years later his funding was terminated, so he decided to give up research and devote full time to teaching. [He then became a well recognized chemical educator, and received the most prestigious ACS Award for Chemical Education.]

It was during this period that I began to lecture in one of the three quarters of freshman general chemistry. I was able to convince my colleagues that I should teach the quarter on "descriptive chemistry." This was the name given to the simple reactions and syntheses using common elements of the periodic table. Lecturers in general chemistry often hated to teach this part of general chemistry. They said the students found it boring, because it was just all memorization of the given equations. I soon became aware that the students were correct, if one followed most of the general chemistry textbooks. These textbooks often devoted sufficient time to this type of descriptive chemistry, but it was usually done in a most boring way. One element (or more) was chosen to be covered in a chapter of the book by discussing the following points: (1) its occurrence on planet earth, (2) preparation of the element, and (3) reactions of the element. The necessary equations to illustrate these points were given in the lecture, and students were then asked to reproduce some of the equations on examinations. No wonder the students were bored with the course and with the memorization of all the meaningless (to them) equations. This aspect of general chemistry is also why many teachers did not like to teach it. This became a catch 22: the instructor did not enjoy teaching the descriptive material; the students were bored by it; which, in turn, resulted in the teacher hating to teach it even more, etc.

After a few years of teaching general chemistry, I began to realize why descriptive chemistry was taking such a rap. I felt that a systematic way to teach elementary chemical reactions and syntheses of "all" the elements, making use of the periodic table, was needed. I realized this would be much more difficult to do than in beginning organic chemistry, where it can be done primarily on the basis of how the functional groups react, be it on a small or a large molecule. Also organic chemistry deals primarily with the elements C, H, O, and N, whereas inorganic chemistry is responsible for all the elements I decided that one systematic approach to elementary inorganic reactions might be to sort them into classes using titles readily understood by beginning students. For example, one class would be a "combination class" which would include two or more substances that combined to form one product, such as copper plus oxygen forms copper oxide ($2Cu + O_2 \rightarrow 2CuO$). Another class would be just the opposite type reaction, "decomposition." Since students know the meaning of these words, they would begin to see that descriptive chemistry can be organized. This becomes even more apparent when the periodic table is used in a meaningful manner to show families of elements and other trends depicted in the table. This approach meant that students could make intelligent guesses of the products of these elementary reactions without having to memorize individual chemical reactions. I emphasized this by telling the students that an examination would never include a reaction that had been written in a lecture or assigned as homework. I wanted them to see the forest (classes) and use this and the periodic table to help them see the trees (individual reactions). I told them if they learned to do this they would be correct 95% of the time which would give them an A in the course. The better students soon learned how to use the periodic table and even enjoyed the challenge.

I wrote a few articles on this approach, which were published in the *Journal of Chemical Education*, and later coauthored a definitive paper on the subject with Professor Robert Parry (**Fig. 3-4**). I feel certain that the development of this method for the teaching of descriptive chemistry was responsible for my receiving the *ACS James Flack Norris Award for Teaching* (1981) and the *ACS George C. Pimental Award in Chemical Education* (1992). The Pimentel Award is the most prestigious award for chemical education given by the ACS, and is not often given to persons

Figure 3-4. Professor Robert Parry, at the University of Utah, and his wife Marj. He started his research on coordination chemistry and then did much work on the main group elements, such as boron (B). His contributions to the importance of science, particularly chemistry, is second to none.

primarily involved in chemical research. However, that is not always true, since Glenn T. Seaborg (1912-99) received the award two years later* and Harry Gray received it this year (2001).

I was also responsible for teaching a junior-senior level course in inorganic chemistry which was the best thing that could have happened to me, despite the amount of work I had to spend on the course. Unfortunately, there was no textbook that I found suitable for the course at that time. I told this to the students and stated that my lectures would emphasize the material that they should learn. This put me on the spot, because for the initial couple of years I had to refer to several different sources to prepare my lectures. The source I found most useful was a book by H. J. Emeleus (**Fig. 3-5**) and J. S. Anderson entitled *Aspects of Modern Inorganic Chemistry*. A few years later, the textbook *Inorganic Chemistry* by Professor Therald Moeller was published. Moeller was on the faculty at the UI and had been teaching such a course for some years. The book made life easier for me and the students.

Figure 3-5. Harry J. Emeleus (1903-93) coauthored the outstanding book *Aspects of Modern Inorganic Chemistry* (1938). His prominent research was on main group chemistry.

One thing all students remembered about my course was that, on the first day of class, they were given a blank periodic table and asked to fill in the elements. They were told that this was a serious test, and that I would collect their papers in twenty minutes. The results were even worse than I expected. It was clear that the students knew very little about how even the most common elements were positioned in the periodic table. They had previously had organic chemistry, but most of them did not even know that carbon is a Group IV element. Some of the students were embarrassed by how poorly they did, but I told them not to worry, because we would make such good use of the table during the course that they would know it without having to memorize it. Years later (1983), as President of the ACS, I had to visit and give talks at many of the ACS Local Sections. Since it was well-known that I enjoyed playing golf, on one such visit, a foursome to play golf had been arranged. After a few holes, one of the golfers said, "You do not remember me but, in 1972, I took your course in inorganic chemistry, and it was the best course I ever had." He told me that part of the reason that he was the

---

*Glenn T. Seaborg and E. M. McMillon were awarded the Nobel Prize (1951) "for their discoveries in the chemistry of the transuranium elements." More recently, an element (no. 106) was named after him, Seaborgium, which was given the symbol Sg. This was the first time that an element had been named after a living person, and it is the highest honor that can be bestowed upon a chemist.

general manager of the regional du Pont plant was because he had made good use of the trends in the periodic table which he had learned in my course. Such remarks always make a teacher feel very pleased, particularly on the occasions when something which was taught in the course is being put to good use.

The reason teaching the junior-senior level course in inorganic chemistry was very important to me was that it put me in direct contact with almost all of the entering graduate students. This was one of the very first courses most of them were required to take. Although it was an undergraduate course, most of our entering graduate students did not have such a course, because many came from a school where junior-senior level inorganic chemistry was not taught. In addition to three lectures a week, there was one synthesis laboratory. This lab gave me a chance to walk around and have a one-on-one talk with the students. Each year, one or two of the students found inorganic chemistry exciting and would choose it for their Ph.D. graduate research. Selwood and I were the only choices they had for director of their dissertation research. This meant that, although most of the students were in organic or physical chemistry, I was able to slowly get some graduate students started in inorganic chemistry.

Every other year I would also teach another course at the graduate level called "Coordination Chemistry." Since my entire research career dealt with this area of chemistry, it was only reasonable that I enjoyed teaching the course. My enthusiasm for the course rubbed off on the students, and we all had a lot of fun together. The lectures were informal, encouraging students to interrupt and ask questions. Thanks largely to the course I had taken from Professor John C. Bailar, Jr., I was able to discuss the early history of metal complexes. Also, as one working in the field, I would at times introduce anecdotal events and history in my lectures. Graduate students, who were required to take two electives, often chose this course because of the stories I was able to tell in addition to the chemistry we had to cover.

For example, the hundred years following Tassaert's discovery of hexaammine cobalt(III) chloride in 1798 left chemists trying to solve the puzzle of the nature of such compounds, which is why they came to be called metal complexes. Worldwide, this occupied chemists doing experiments, proposing concepts, hypotheses, and even theories to account for the existence of such compounds. These were all proven to be wrong, and chemists were left with two theories—one called the *chain theory* proposed by S. M. Jørgensen and the other the *coordination theory* proposed by Alfred Werner in 1893. Jørgensen was much older than Werner, and Jørgensen had prepared many of the metal complexes by making use of his chain theory. He believed his theory to be correct and, in addition, was much better at the preparation of such compounds than was Werner. It upset Jørgensen very much to have Werner, a young organic chemist, challenge his chain theory. In fact, the sabbatical year I spent in Copenhagen I spent doing research in the old lab that was used by Jørgensen. I was told many stories about the polemic differences between the two, but there is one I liked best.

Niels Bjerrum, father of my host Jannik and one of Jørgensen's students, told me that his students and assistants were not allowed to talk about Werner's theory. However, at times, they would lock themselves in the men's room and discuss the merits of each theory in the absence of Jørgensen. He finally was the one to prepare the compound [Ir(NH$_3$)$_3$Cl$_3$], predicted by the coordination theory to be a molecular nonelectrolyte, whereas his chain theory predicted it would be an ionic electrolyte. This observation by Jørgensen proved his theory was wrong, making this a very sad day for him as an old man at the end of his career as an outstanding synthetic inorganic chemist. Jørgensen then did what all creditable scientists would do: he published the results of his experiments, proving his theory wrong. Werner's coordination theory, proposed in 1893, almost 100 years since the discovery of metal complexes, has withstood the test of time and would appear to be correct. At present, most of the chemistry of metals involves coordination chemistry as in bioinorganic chemistry, new solid state materials, and organometallic compounds. Werner was the first chemist in Switzerland to receive the Nobel Prize, which he obtained in 1913 for his theory. He and Jørgensen had never met but after Werner received his Prize in Stockholm, he stopped in Copenhagen to see Jørgensen who was in the hospital near death and did not see Werner.

It was slow going for a time but, if you read on, you will see that inorganic chemistry at NU has become one of the strongest in the US, and now ranks among the best in the world. This year, *US News and World Report* placed NU inorganic chemistry as one of the top four in the nation (others being the University of California, Berkeley, Caltech, and MIT).

It has delighted me, over the past 55 years, to see our department grow in strength in inorganic chemistry. This was due to our successful hires of first rate faculty (see pg. 79) in different and important areas of research. Also important has been the friendliness among faculty and their research groups. This results in free discussions among the groups that often result in joint work and publications.

RESEARCH

Research Environment

When the six of us joined the chemistry faculty at NU in 1945-47, universities and departments provided almost no start-up funds to assist beginning faculty with their research needs. Also, there were only a few funding agencies to which a new faculty member could turn for financial help. One of these was the *Research Cooperation* (RC), which funded successful proposals with about $2,000 for one or two years. Fortunately, for the new faculty, the University had also received some discretionary funds from the Abbott Foundation, which were distributed among various departments. The chemistry department made good use of its Abbott funds. The small amount it had was used entirely for the support of research. This made it possible for beginning faculty to set up a relatively adequate lab.

The chemistry department made every effort to help new faculty get a good start on their research. For example, glassware and chemicals needed for research were often purchased with funds the university allocated to the department for undergraduate teaching laboratories. Also, the teaching loads of beginning faculty were kept as low as possible. In most cases, for the first few years, the new faculty did not give lectures, but were responsible for small discussion groups and their labs, as noted earlier.

Being interested in the stereochemistry of transition metal complexes, I had a need for a spectraphotometer and a polarimeter. Fortunately, Irving Klotz gave me one of his Beckman spectraphotometers, and Robert Baker (1908-92) gave me his Zeise polarimeter. Ordinary chemicals, such as cobalt salts, were readily available, but precious metals, such as platinum, were very expensive. Research with these costly metals required the complete recycling of reaction wastes and products so that the metals could be used over and over again.

Funding for research slowly improved, followed by a sudden increase after the Russian launching of *Sputnik* in 1957. The establishment of the National Science Foundation (NSF) took place in 1950. The first NSF grants to our department were awarded to Robert Burwell, Arthur Frost, and myself in 1952. My grant was for $16,100, over a two year period. At about the same time, Ralph Pearson and I (**Fig. 3-6**) received a grant of $7,000 from the Atomic Energy Commission (now Department of Energy) to support our joint project. While minuscule by today's standards, this was a princely sum in those days. As pleased as we were with the money, we were even more pleased by the fact that the government was finally aware of the importance of supporting basic research.*

Figure 3-6. Ralph G. Pearson of "hard" and "soft" fame, on the left, and me during our many years (1950-73) of collaborating on research, coauthoring 60 publications, and writing our book on mechanisms of inorganic reactions

---

*This period of the early 1950s was followed by continued support of long range fundamental research and prompted my colleague Malcolm Dole (1903-90) to title his autobiography in 1989 *My Life in The Golden Age of America*. This title is most appropriate, particularly now with the rapidly expanding technological developments. We scientists know that the outburst of technology of such proportion could not have occurred were it not for prior significant basic research. This can be illustrated with but one of many examples. As an inorganic chemist, my choice of illustration is that of the silicon chip which makes much of our present technology possible. Basic research by inorganic chemists established the fundamentals of how to produce silicon from its oxide or ordinary sand. Other scientists could, of course, provide other examples. It is important for scientists to continually deliver this message to

Figure 3-7. NU Inorganic Chemistry Faculty, 2001. Left to right: (back row) Ken Poeppelmeier, Du Shriver, Jim Ibers, Tom O'Halloran; (front row) Lou Allred, Tobin Marks, Fred Basolo, Chad Mirkin, Hilary Arnold Godwin. Credit: © Evanston Photographic Studios, Inc.

During the late 50s and early 60s it was apparent that the federal government had become aware of the need for the support of long range basic research in our universities. Not only was NSF established and provided with increasing funds for basic research, but other agencies were also receiving additional funding. For example, the National Institutes of Health (NIH) began to get enough research funds to enable it to direct some of it toward chemistry. In fact, during a period when it seemed there was going to be a shortage of chemists, NIH had a fellowship program to support graduate students in chemistry. Beginning with Ralph Pearson and myself, inorganic chemistry at NU has benefitted from this support. Over the years we have continued to add other inorganic chemists to our departmental faculty: Louis Allred (1956), Duward Shriver (1961), James Ibers (1964), Tobin Marks (1970), Kenneth Poeppelmeier (1984), Thomas O'Halloran (1986), Chad Mirkin (1991), and Hilary Arnold Godwin (1996) (**Fig. 3-7**). All of these inorganic chemists are doing research of the highest quality, which enables them to receive excellent funding and to publish in the best scientific journals. During 1999 this group of inorganic chemists published a total of 71 articles on their original

---

funding agencies: support of long range basic research is needed if the US is to keep ahead of other countries in the rapid growth of the global technological industry. Likewise important is the message to be emphatically delivered to the general public that problems of basic research can be very complicated and require years to find the desired solutions. Witness the years of research to find a cure for cancer and that the breakthrough may now be at hand using gene therapy.

research. In addition, the following faculty members of the department, who teach and do research in organic and in physical chemistry, also make major contributions to inorganic chemistry as some of their research is related to inorganic systems: Irving Klotz, Joseph Lambert, Brian Hoffman, Joseph Hupp, and SonBinh Nguyen.

During my years of research activity at NU (1946-92), I was awarded funding by RC, NSF, NIH, AEC (Atomic Energy Commission, now the Department of Energy), and AFOSR (Air Force Office of Scientific Research). During all of these years, to and beyond the present, the government has allocated hundreds of billions of dollars for the support of long range scientific research. It is increasingly important that this be continued and even increased because industry, in its effort to cut costs, is no longer doing long range research at any appreciable level. It is much less expensive for industry to have this type of research done in universities by faculty members and their students.

Research at NU

Since there were no graduate students at NU interested in choosing inorganic chemistry for their Ph.D. dissertation research, I was the only author on my first paper, which was published in *The Journal of the American Chemical Society* (*J. Am. Chem. Soc./JACS*). Publication number two (*JACS*) was on my Ph.D. research done with my mentor, Professor Bailar, at UI. Publication number three (*JACS*) was also authored by me alone, and again it had to do with the syntheses and spectra of cobalt(III) complexes.

Figure 3-8. Sir Christopher K. Ingold (1893-1970), a first rate physical/organic chemist. Among his many contributions, he, along with Hughes, gave us the symbols S_N1 and S_N2 for nucleophillic, mononuclear, and binuclear reactions, respectively.

As indicated earlier, at Rohm and Haas I had become interested in studies of physical organic chemists on the kinetics and mechanisms of substitution reactions on carbon. In fact, Hughes and Ingold (**Fig. 3-8**) had classified such reactions as S_N1 for unimolecular nucleophilic substitution mechanisms and S_N2 for bimolecular nucleophilic substitution mechanisms. It occurred to me that studies of this type could also be carried out with metal coordination complexes. Because of my experience with cobalt(III) and platinum(II) complexes, I thought this would be a good place to start. However, I did not have any experience with investigations on the kinetics and mechanisms of chemical reactions. Fortunately, among our group of newly arrived instructors there was one person making such studies on organic compounds. This was Ralph G. Pearson, who was collaborating with our older colleagues Carroll King and

Faculty Position at Northwestern University

Figure 3-9. Attendees at a symposium on the Kinetics and Mechanisms of Inorganic Reactions at NU in 1957.

Fredrick Bordwell. As I saw it, my job would be to convince Ralph that he should come aboard and help me do research of the same type on inorganic compounds. This was not easy and it required a couple of years of repeatedly bringing it up to him. Ralph's initial response was, "There is no interest in inorganic chemistry, so why should I waste my time studying such systems?" He pointed to the fact that we did not even have any graduate students expressing an interest in inorganic chemistry.

Ralph was correct at the time, but things changed dramatically about then, with inorganic chemistry slowly becoming increasingly important. I was finally able to seduce him into helping us with the kinetics and mechanisms of our research, and our first joint paper was published in 1952 in *JACS*. In fact, he became so interested in our effort together that he later became considered a physical inorganic chemist. For his work over the years he received the *ACS Award for Distinguished Service in Inorganic Chemistry*.

Ralph and I, with help from our excellent students, joined forces to get inorganic chemistry started at NU. He and I published 60 scientific papers jointly at a time when it was just beginning to receive attention in the US. Because of this, we were in an enviable position to attract some of the very best graduate students, such as Andy Wojcicki, Harry Gray, Bob Angelici, Earl Thorsteinson, Don Morris, Ken Raymond, Al Crumbliss, Ruth Kowaleski, and many others (see Appendix for names of graduate students and postdoctorates). Ralph and I also wrote a book entitled *Mechanisms of Inorganic Reactions*, published by John Wiley & Sons in 1958 with a second edition in 1967. In order to focus attention on the book and because it had been eight years since the symposium at the University of Chicago, we sponsored a symposium at NU in 1957 (**Fig. 3-9**). The book was received extremely well worldwide. It produced outstanding reviews, with one of the reviewers, Daryle Busch, stating that it was destined to become the "bible of inorganic mechanisms."

Unfortunately, Pearson left NU in 1976 to accept a position at the University of California, Santa Barbara. He would have preferred to stay at NU, but his wife was ill and having a difficult time with the Chicago winters. Ralph soon became best known for his publications on the *Hard and Soft Acid-Base Concept*. The concept is very qualitative and descriptive but, keeping that in mind, it works. It is now in all beginning inorganic textbooks. When Ralph first mentioned the concept to me, my reaction was, so what is new? Coordination chemists have known for years that some metals coordinate better with oxygen ligand atoms (M-OR), whereas others prefer sulfur ligand atoms (M-SR). Even as far back as the 19th century, Berzelius (1779-1848) made the point that certain metals occur on the earth's crust as oxides and others as sulfides. Ralph did not take my advice and pressed on with this, until now the concept is even often used in research papers by chemists and molecular biologists. This shows how important it is to assign a common name to a useful scientific phenomenon.

As mentioned above, NU has always encouraged strong interaction among its faculty, which has contributed to its uniqueness. It is also unique, perhaps, because several instructors arrived at about the same time. We had paper bag lunches in one of our offices each day, where we talked about various things in the news media but, at times, also about our chemistry. We soon became good personal friends and often collaborators in research. At some point, the older faculty wondered what we young turks might be up to, so they suggested joining us and using a conference room for our lunch. This location continues, and it is important to facilitate a close friendly relation among all of the faculty. Once a week a member of the faculty gives a talk on their research, which at times results in a joint publication with one of his colleagues. Our department has always had a chairperson, rather than a head. Departmental meetings are held to decide on its needs and on important policy changes. At these meetings everyone can have their say, with one vote counting the same whether it be from an instructor or from a distinguished professor. As a result, the university administrators know that when our department makes a request, it comes from the entire faculty, so they do as much as possible to grant the department's request. Consequently, our department was able to add needed faculty, and to keep faculty who obtained offers from other outstanding schools.

Figure 3-10. Bradley Holliday, a graduate student, giving a blackboard account of his research in progress during the Saturday morning BIP (Basolo-Ibers-Pearson) meeting. These meetings are long standing within the inorganic chemistry group—faculty, postdoctorates, and graduate students participate voluntarily.

A good example of the strength of the interactions between inorganic research groups is the Saturday BIP group (Basolo-Ibers-Pearson) (**Fig. 3-10**). The acronym BIP was coined by Ken Raymond, who was a graduate student when the Saturday groups that met together were those of Ibers, Pearson, and myself. These meetings started when Ralph and I were collaborating on much of our research. We had an informal gathering once a week where one or two group members (graduate students or postdoctorates) were asked to present a blackboard chalk talk on their research in progress. No slides were used and the meetings were not like seminars where one reports on their generally finished and successful results.

In fact, the most interesting and useful presentations were those giving unsuccessful results, because this caused the greatest amount of discussion as to what might be going wrong. It also gave me a chance to win many 25¢ bets over the years. The students would make a statement that a certain result would take place, and I would bet them that it would not. They never realized that they only had one chance of their guess being correct when any of three or four things might happen. This meant that I had two or three times the chances to their one chance. One of our students, Steve Strauss, lost his bet and gave me a check for 25¢. I never did cash it, so he was not able to balance his checking account. On my 70th birthday celebration, he presented me with a plaque consisting of a newly-minted quarter and the statement: "For his lucky winning of some of his 25¢ bets."

Since Ralph and I could not find a convenient time to meet without a class conflict, we decided to meet at 9:00 AM on Saturdays. We chose this time because it would allow us to get to the football stadium in time for the kick-off of yet another game that NU would lose. Saturday was also chosen in order to try to convince members of the research groups that Saturday, too, might be a work day, but we had little success with this. Ibers and his group asked to join us in order to give them some exposure to inorganic syntheses and reactions. Up to that point, Jim was one of the world's experts on structural inorganic chemistry, mostly using X-rays. Soon after that, all of the inorganic chemists at NU started to attend and take part in the presentations and discussions. There is no obligation to attend BIP and no class credit is received, but almost all of the inorganic chemists come each week. This means our conference room is filled with some 40 people or more. We believe this gives students and postdoctorates the experience needed to present a good, relaxed informal discussion of their research, and to handle unexpected questions that may be asked. In fact, some of our alumni tell us that the most helpful thing in our graduate program is BIP.

Except for some anecdotal comments, what follows may only be of interest to chemists. Others may choose to omit reading these sections and go directly to the final portion of the book. These sections on specific research will present only some of the work done by students, postdoctorates, senior collaborators, and myself. (See Appendix for names. To all of these people, I owe my sincere thanks.)

Ligand Substitution Reactions of Octahedral Cobalt(III) Complexes (Co(III))

Since my experience with metal complexes stemmed from my dissertation research with Professor Bailar who was mostly interested in stereochemistry, my first approach to the study of their reaction mechanisms was to investigate the stereochemical changes during reaction. Before tackling this, I read a few of Alfred Werner's papers (translated into English by George Kauffman), attempting to get meaningful information on the stereochemical changes accompanying ligand substitution reactions of some cobalt(III) complexes. Since there were no spectrophotometers at the time, Werner and his students had to depend on the visual changes in color during the reactions. With the advantage of having a spectrophotometer, we wanted to get more reliable qualitative data and perhaps even quantitative data on the reactions studied by Werner, as well as other reactions.

My graduate student, Bob Stone, discovered that during a ligand substitution reaction, generally, a mixture of stereoisomeric products are obtained. A look at molecular models of octahedral complexes show that different product structures could result from different mechanisms of reactions. We thought we would use the space-filling Fischer-Hershfelter model kits available for the use of organic chemists to aid in understanding the stereochemical changes. However, the kit did not contain an octahedral structure, which was needed by coordination chemists. I pointed this out to a Fischer employee and he not only had the octahedral model added to the kit, but also added five-coordinate and square planar four-coordinate models. Later, coordination chemists would thank me for having gotten these structures added to the kit, which helped make it useful for inorganic chemists. At that time I also wrote a review article (*Chem. Revs.*, **52**, 459 (1953)) to show the possible structures of products that may be obtained by ligand substitution reactions of octahedral metal complexes (see pg. 85). Again, both models and experiments showed that stereochemical changes accompanying substitution reactions of octahedral complexes, such as Co(III) compounds, are much more complicated than are those of tetrahedral compounds, such as at carbon in organic compounds.

In 1953, Pearson and I published a paper in *JACS,* with our graduate student Bob Stone, on the stereochemical changes during reactions of some Co(III) complexes. In addition, we were investigating the kinetics of these reactions. Independent of our kinetic studies, similar studies were being done at University College London by the research groups of Sir Christopher Ingold (a first rate physical organic chemist) and Sir Ronald Nyholm (an outstanding coordination chemist). However, they were not investigating stereochemical changes of the reactions, but only the kinetics. When they needed stereochemical information, they referred to our paper. I have been told that Ingold would often ask to see the "Bob, Fred, Ralph paper" during research group discussions. Ingold was very British and did this to make fun of how Americans use trivial names such as Bob instead of Robert and Fred instead of Fredrick or Alfred.

Dissociation Process (S_N1) for *trans*-[M(AA)$_3$ax]

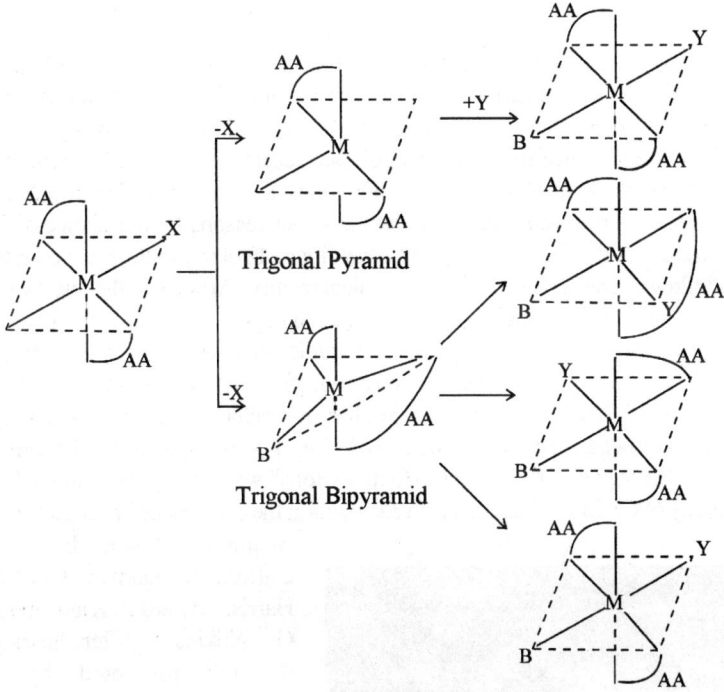

Dissociation Process (S_N2) for *trans*-[M(AA)$_3$ax]

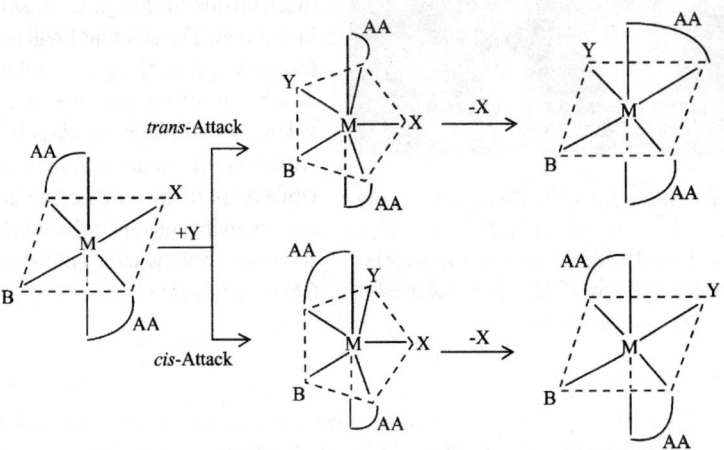

Acid Hydrolysis or Aquation of Metal Ammine Complexes of Cobalt(III)

The earlier work mentioned above on the detailed kinetic studies could not have been done without the joint effort of Pearson and myself. My contribution was primarily to select and synthesize the metal complexes to study, and Pearson was responsible for the kinetic studies. Since most of the published research made use of stable Co(III) compounds, we decided to use these compounds to investigate ligand substitution reactions of six-coordinated metal complexes. Other reasons for our choice of Co(III) ammines were that they react slowly and have different colors, so the reaction rates can easily be followed spectroscopically by classical means. Articles in the late 1940s and early 1950s dealt mostly with the syntheses and characterization of metal complexes. Of course, there were some reports of kinetic studies done primarily by J. N. Brønsted, F. J. Garrick, and A. B. Lamb, but these were concerned with salt effects on the behavior of ionic reactants. However, interest in studies of mechanistic details of reactions of metal complexes began to appear in the early 1950s. For example, in 1950 a conference on "Rate and Equilibrium Behavior of Complex Ions" was arranged by Henry Taube at the University of Chicago (**Fig. 3-11**). Participants at the conference included Arthur W. Adamson, Robert E. Connick, Clifford S. Garner, Gordon M. Harris, Edward L. King, and Ralph G. Wilkins. After hearing the research presented by these outstanding investigators, it was clear that there was about to become a serious focus on the mechanisms of inorganic reactions in solution. The talk that I will never forget was that given by Taube, in which he discussed the use of valence bond theory to classify the reactivity of metal complexes as labile or inert. In 1952 he published his reactivity concept, which I think is the best review ever published in *Chemical Reviews.**

Figure 3-11. Professor Henry Taube (Nobel Prize, 1983) and me when he received the Robert A. Welch Award. I was President of the ACS and was there as its representative. Also in 1983, Taube received the NAS Award in Chemical Science.

---

*Taube (Nobel Prize 1983) was able to use the valence bond theory to assign qualitative rates of ligand substitution of metal complexes. He considered only six-coordinate octahedral complexes, which are, by far, the most common. His classification was that inner orbital complexes are substitution inert, whereas outer orbital complexes are labile. With this information he was able to design a test tube-like experiment which won him the Nobel Prize

Our initial publication in *JACS* on reaction mechanisms appeared in 1952 and was entitled *Dissociation Mechanism for the Aquation of Some Cobalt(III) Complex Ions*. This research was done by Charlie Boston. An aquation reaction, as shown (eq. 3-1), is one where water replaces one or more ligands of the metal complex.

$$[Co(NH_3)_5Cl]^{2+} + H_2O \xrightarrow{slow} [Co(NH_3)_5H_2O]^{3+} + Cl^- \quad (3\text{-}1)$$

The NH_3 ligands were replaced by alkyl amines of increasing size, and it was found that the rates of reaction increased with increasing size of the ligands. This result was best explained by a rate-determining dissociative process, which makes room for the larger ligands. If this involved a bimolecular mechanism, the rates would be expected to decrease with increase in ligand size, as it would be more difficult for the entering nucleophile to attack the metal. In terms of the Hughes Ingold notation of S_N1 for unimolecular and S_N2 for bimolecular reactions, the aquation reactions of these Co(III) ammines are believed to involve a S_N1 mechanism (eq. 3-2).

$$[Co(NH_3)_5Cl]^{2+} \xrightarrow{slow} [Co(NH_3)_5]^{3+} + Cl$$

$$\text{fast} \downarrow H_2O \quad (3\text{-}2)$$

$$[Co(NH_3)_5H_2O]^{3+}$$

Since the solvent in these experiments was water, the rate dependence on the concentration of water could not be determined. Thus, it was necessary to get information on the mechanism by using an indirect approach, such as the effect of increasing the size of the innocent amine ligands.

The experiments were based on his knowledge that Co(III) and Cr(III) complexes are substitution inert, whereas Co(II) and Cr(II) are labile. Thus, if he were to mix a Co(III) with a Cr(II), reaction, (3-3) would take place.

$$[Co(III)(NH_3)_5{}^{36}Cl]^{2+} + [Cr(II)(H_2O)_6]^{2+} \xrightarrow{HCl} [(NH_3)_5Co-{}^{36}Cl-Cr(H_2O)_5]^{4+}$$

$$\rightarrow [Co(II)(H_2O)_6]^{2+} + [Cr(III)(H_2O)_5{}^{36}Cl]^{2+} + 5NH_4Cl. \quad (3\text{-}3)$$

This involves the formation of the activated bridged intermediate $[(NH_3)_5Co-{}^{36}Cl-Cr(H_2O)_5]^{4+}$ and is called *inner sphere electron transfer*. If the Cr(II) complex is $[Cr(bipy)_3]^{2+}$, then the bridged intermediate cannot be formed, because of the bulky bipy ligand, and the reaction proceeds by an *outer sphere electron transfer* process. This is chemistry that even I understand, and it shows the clear thinking of Taube that I admire.

Base Hydrolysis of Metal Ammine Complexes

It had long been known that the base hydrolyses (eq. 3-4) of such complexes are orders of magnitude faster than their acid hydrolyses. However, the mechanism of base

$$[Co(NH_3)_5Cl]^{2+} + OH^- \xrightarrow{fast} [Co(NH_3)_5OH]^{2+} + Cl^- \qquad (3\text{-}4)$$

hydrolysis had not been investigated. It was generally believed that this was due to OH^- being a much stronger nucleophile than H_2O and would therefore have a much greater tendency to attack the metal. However, our experiments had suggested that the acid hydrolysis of the Co(III) complex did not occur by a nucleophilic attack of water, but by a dissociative process. For this reason we were skeptical about the assignment of an S_N2 mechanism to these reactions. Students and postdoctorates who worked on the base hydrolysis were John Bergman, Bob Meeker, Pat Henry, Ron Munson, Dr. Hans-Herbert Schmidtke, and Professor Art Adamson.

The contrary view was taken by Sir Christopher K. Ingold and Sir Ronald S. Nyholm of the University College, London. They published papers supporting a S_N2 base hydrolysis reaction, and stated that our S_N1CB (substitution, nucleophilic, unimolecular, conjugate base) mechanism (eqs. 3-5, 3-6, and 3-7) was not correct.

$$[Co(NH_3)_5Cl]^{2+} + OH^- \xrightleftharpoons{fast} [Co(NH_3)_4(NH_2)Cl]^+ + H_2O \qquad (3\text{-}5)$$

$$[Co(NH_3)_4(NH_2)Cl]^+ \xrightarrow{slow} [Co(NH_3)_4)(NH_2)]^{2+} + Cl^- \qquad (3\text{-}6)$$

$$[Co(NH_3)_4(NH_2)]^{2+} + H_2O \xrightarrow{fast} [Co(NH_3)_5OH]^{2+} \qquad (3\text{-}7)$$

Garrick had reported the rapid H/D exchange of $[CO(NH_3)_6]^{3+}$. Pearson and I began to think that this mechanism might be responsible for the rapid reactions of metal ammine complexes with OH^-. Since reaction mechanisms are scientific theories which can only be proven wrong, we had considerable work ahead of us to test our S_N1CB mechanism. One thing that needed to be done was to determine if a complex devoid of ligands containing N-H bonds, such as $[Co(py)_4Cl_2]^{1+}$ (py is pyridine with no N-H bonds), would also react very rapidly with OH^-. Our mechanism required that the reaction rate of the pyridine Co(III) complex be almost independent of the pH of the solution. And this is what we found, which was consistent with the S_N1CB process.

Another test of this mechanism was to determine if the rate of H/D exchange was faster than the rate of base hydrolysis, as was required by our mechanism. We found the rate of exchange to be significantly faster than the rate of base hydrolysis, as required by the S_N1CB mechanism. These exchange experiments were hands-on experiments done by Arthur Adamson (**Fig. 3-12**) and me. We both happened to be in Copenhagen on sabbatical during the academic year 1954-55. We did our research in the same laboratory in the old University of Copenhagen. This was a lab with a good history, having been used by Brønsted of acid-base fame, and S. M. Jørgensen for his synthesis of most of the metal complexes which inspired A. Werner to give us the coordination theory. Now coordination chemistry, which involves metals that are bonded to groups called ligands, is implicated in most of the chemistry of metals.

One other thing of interest is how Arthur and I measured the H/D exchange. We were invited by Lunderstrom-Lang, director of the Carlesberg Lab, to use his gravity apparatus. [Carlesberg is a very fine beer so there is no need to do beer research to improve it.] This meant that Lunderstrom-Lang, a biochemist, was doing protein research. When we arrived at his lab, he did not let us touch his apparatus because he said it was too sensitive and he would rather we not use it. We prepared the solutions and he, with an assistant, made the H/D measurements. As it turned out, we spent a worthwhile day on our research, and also had the pleasure of listening to Lunderstrom-Lang tell us many different and entertaining stories.

Figure 3-12. Professor Arthur Adamson at the University of Southern California. Art is extremely well known by coordination chemists for his seminal work on the photochemical reactions of metal complexes. In addition, his research on surface chemistry is highly regarded by his peers.

Ingold was acknowledged as the high priest of organic mechanisms whom, it is said, should have obtained the Nobel Prize. He already had worked out the rules for nucleophilic substitution reactions at tetrahedral carbon, and was prepared to move on to new geometries and coordination numbers—this meant investigations of metal complexes, giving particular attention to our work on substitution reactions of Co(III) complexes. We had published the experiments done to test the validity of our proposed S_N1CB mechanism, and found that each of the steps in the mechanism withstood the tests. However, Ingold was able to give a different interpretation to our observations. Since one cannot prove that a mechanism is correct but only wrong, we decided to try to prove that Ingold's proposal of an S_N2 reaction was wrong. This proof was finally obtained by Hans-Herbert Schmidtke, our postdoctoral who conducted a study of the reaction of trans-$[Coen_2NO_2Cl]^+$ with OH^- and NO_2^- in dimethylsulfoxide as solvent. The results obtained are shown by equations (eqs. 3-8 through 3-11).

$$[\text{Co en}_2\text{NO}_2\text{Cl}]^+ + \text{NO}_2^- \xrightarrow{\text{slow}} [\text{Co en}_2)\text{NO}_2)_2]^+ + \text{Cl}^- \qquad (3\text{-}8)$$

$$[\text{Co en}_2\text{NO}_2\text{Cl}]^+ + \text{OH}^- \xrightarrow[\text{fast}]{\text{very}} [\text{Co en}_2\text{NO}_2\text{OH}]^+ + \text{Cl}^- \qquad (3\text{-}9)$$

$$[\text{Co en}_2\text{NO}_2\text{Cl}]^+ + \text{NO}_2^- + \text{OH}^- \xrightarrow{\text{fast}} [\text{Co en}_2(\text{NO}_2)_2]^+ + \text{Cl}^- + \text{OH}^- \qquad (3\text{-}10)$$

$$[\text{Co en}_2\text{NO}_2\text{OH}]^+ + \text{NO}_2^- \xrightarrow[\text{slow}]{\text{very}} [\text{Co en}_2(\text{NO}_2)_2]^+ + \text{OH}^- \qquad (3\text{-}11)$$

These reactions definitely show that, although reaction (3-9) is very fast, the competition (reaction 3-10) between NO_2^- and OH^- for reaction favors NO_2^-. This shows that OH^- had not attacked Co(III) to form the hydroxo compound (3-9), the product of reaction would have been $[\text{Co en}_2\text{NO}_2\text{OH}]^+$ and not $[\text{Co en}_2(\text{NO}_2)_2]^+$, because reaction (3-11) is very slow. Therefore, the results of these reactions afford proof positive that the base hydrolysis does not proceed by a S_N2 mechanism.

In 1962, I was invited to present a paper at a meeting of the Royal Society of Chemistry in London. At that time I was also invited to give a seminar and spend the day at University College London. After my talk, Ingold invited me to his office where we had a friendly and interesting discussion about our different views on the substitution reactions of Co(III) complexes. In my talk I had not mentioned the experiments we did that proved his S_N2 mechanism wrong, so I told him about this. He thought about it for awhile, and then said, "Yes, but dimethylsulfoxide as a solvent is not water." My response was, "That is true, but I have recently been told that some yet unpublished work of Taube using isotopically labeled oxygen also showed the S_N2 mechanism in water to be incorrect." Having related this to Ingold, we then went on to talk about other research that we had in progress in our respective groups. I came away believing that Ingold was a very fine British gentleman, even if one does not get that impression from reading his articles.

Shortly after this polemic, others worldwide carried out competition experiments, all of which supported our position. Even Martin Tobe, who worked on the problem at University College London, finally agreed with us, but Ingold never changed his position in the literature. However, the very fact that he interested himself in the subject of inorganic reaction mechanisms added great stature to the topic. Both Pearson and I recall the sparse audiences to which we would lecture prior to 1953, and the crowds (many organic chemists) we drew after Ingold's entry into the field.

Pearson and I were given the faculty title of Instructor, to be followed by Assistant Professor, and Associate Professor with tenure. Had our polemic with Ingold and Nyholm, two of the giants of chemistry, not turned out in our favor, we may not have gotten tenure and our department may not now rank among the top four in inorganic chemistry in the U.S. We were very fortunate to have this exchange focus so much favorable attention on our department just at the time when inorganic chemistry in the US was beginning to achieve the importance it now has.

The amount of time that I have spent discussing the base hydrolysis of Co(III) complexes was done for two reasons—one being that it attracted the most attention of all the 60 publications co-authored by Pearson and myself, thanks in large part to Ingold and Nyholm. The other reason is to caution young chemists who take on a giant of chemistry that they had better be right or they may not get tenure.

Linkage Isomers

My interest in the possible syntheses of linkage isomers stemmed from the knowledge that we and others were beginning to accumulate on the mechanisms of reactions of metal complexes. I believed that we were in a position to make use of the mechanisms to design the syntheses of new metal complexes, or of known complexes by new methods. In 1961, I published a feature article in *Chemical & Engineering News* which provided examples of the methods of the organic chemists being applied to the syntheses of metal complexes. One of the examples was a result of a study we made on the apparently simple reaction shown by equation (3-12).

$$[Co(NH_3)_5Cl]^{2+} + NO_2^- \xrightarrow{\Delta} [Co(NH_3)_5NO_2]^{2+} + Cl^- \quad (3\text{-}12)$$

A detailed kinetic study done by John Bergman and Pat Henry demonstrated this to be anything but a simple reaction. The kinetic data showed the overall reaction involved five steps:

$$[Co(NH_3)_5Cl]^{2+} + H_2O \longrightarrow [Co(NH_3)_5OH_2]^{3+} + Cl^- \quad (3\text{-}13)$$

$$[Co(NH_3)_5OH_2]^{3+} + H_2O \longrightarrow [Co(NH_3)_5OH]^{2+} + H_3O^+ \quad (3\text{-}14)$$

$$2HNO_2 \longrightarrow N_2O_3 + H_2O \quad (3\text{-}15)$$

$$[(NH_3)_5Co\text{—}OH]^{2+} \xrightarrow{N_2O_3} \begin{bmatrix} (NH_3)_5Co\text{—}O\text{---}H \\ | \\ O\text{—}N\text{---}NO_2 \end{bmatrix}^{2+} \quad (3\text{-}16)$$

$$\xrightarrow{\text{slow}} [NH_3)_5Co\text{—}ONO]^{2+} + HNO_2 \quad (3\text{-}17)$$

$$[(NH_3)_5Co\text{—}ONO]^{2+} \xrightarrow{\text{very slow}} [(NH_3)_5Co\text{—}NO_2]^{2+} \quad (3\text{-}18)$$

Of particular significance in this reaction scheme was the O-nitrosation step (eq. 3-16), which suggested that the Co-O bond was not cleaved and that the kinetic product was therefore the unstable nitrito (Co-ONO) isomer (eq. 3-18). This then rearranged to form the stable nitro (Co-NO$_2$) linkage isomer. Oxygen-18 experiments by Taube and his postdoctorate, Dr. Kent Murmann (my Ph.D. student), subsequently confirmed the fact that there is no rupture of the Co-O bond during this process.

In the early sixties, women rarely went on to do graduate work in the sciences and engineering. Our department had only six women, two of whom were in my research group. After talking to Geneva Hammaker, one of my two female graduate students, about research problems of interest to me, she chose to prepare the nitro/nitrito isomers of Rh(III) and Ir(III). Her plan was to start with the [M(NH$_3$)$_5$Cl]$^{2+}$ compounds, as had been used for the preparation of the Co(III) isomers. First, she had to prepare the [M(NH$_3$)$_5$Cl]Cl$_2$ starting compounds. She went to the library to get the information needed to prepare these compounds. Having done that, she came to tell me what she found.

She stated that she found a paper by S. M. Jørgensen, in which he reported his attempts to prepare these isomers of the Rh(III) complex. He failed, always only obtaining the nitro compound.[*] I asked her, "What were the experimental conditions?" She said, "Boiling water for some hours." By this time in our conversation, she had begun to cry, saying, "How can I be expected to make these isomers if Jørgensen failed?" I told her that he was an outstanding experimentalist who knew that the ordinary substitution reactions of Rh(III) and Ir(III) were very slow and required heating at prolonged times to occur. Under these conditions, he always obtained the thermodynamically stable nitro complex. I said, "Why not try the reactions at low temperature, because the mechanism we expect does not require breaking of the M-O bond to yield a kinetic product of the nitrito complex M-ONO." Were this to take place by M-O cleavage, then Jørgensen would be correct and the Rh(III) and Ir(III) compound would require elevated temperatures to react." I even suggested trying the reaction in an ice bath. She finally agreed that this should be attempted. Some days later she came to see me, all smiles, very happy to report that she was the first to isolate the nitrito complexes of Rh(III) and Ir(III), as well as the first to measure their rates of isomerization to the stable nitro forms. Later, when we published her research, she again became ecstatic and honored (as was I) to see her paper appear as the very first article in *Inorganic Chemistry*. The reference, *Inorg. Chem.*, **1**, 1, (1962), is one that I have, therefore, always found easy to remember. I owe this to my good friend and professional brother

---

[*]The nitrito/nitro isomers of Co(III) were prepared by S. M. Jørgensen in the late 19th century. His laboratory work was truly outstanding, and he prepared and did the wet analysis of his new compounds himself. These compounds were on display in the laboratory where he did his research using only a Bunsen burner and classical glassware. He was guided in his syntheses by his chain theory which he later proved to be wrong. He also tried to prepare the nitro/nitrito isomers of the other two members of the Co(III) triad, Rh(III) and Ir(III), but he was only able to prepare the stable nitro compounds.

R. Parry (who also had J. C. Bailar, Jr. as his mentor). Bob was the journal's first editor, and the person most responsible for getting it off to its excellent start. *Inorganic Chemistry* is now one of the very best journals, committed to articles on inorganic chemistry.

Having had this success with making new linkage isomers, John Burmeister and I decided to have another go with our approach of using mechanisms to design methods of synthesis. It was well known that the ligand NCS⁻ formed both thiocyanato (M–SCN) and isothiocyanato (M–NCS) complexes, depending on the central metal. However, no linkage isomers had been observed. One attempt was made by the redox reaction between $[Co(NH_3)_5NCS]^{2+}$ and $[Cr(H_2O)_6]^{2+}$ to give the bridged species $[(H_3N)_5Co-NCS-Cr(H_2O)_5]^{4+}$. This bridged system may then cleave to yield $[Cr^{III}(H_2O)_5SCN]^{2+}$ as the kinetic product which would rearrange to the thermodynamically stable $[Cr^{III}(H_2O)_5NCS]^{2+}$. This approach was tried many times using different conditions, but with no success in isolating the unstable thiocyanato complex.

Figure 3-13. John L. Burmeister, Alumni Distinguished Professor and Associate Chairman, Department of Chemistry and Biochemistry, University of Delaware. He prepared the first linkage isomers of metal complexes of the type M–NCS and M–SCN.

During the academic year 1961-62, I went on a sabbatical leave to the University of Rome, which was discussed near the end of Chapter 2. There, one day as I was browsing through some journals in the library, I came across an interesting article by A. Turco and C. Pecile, in which they reported a change in NCS⁻ bonding to platinum(II), depending on the other ligands coordinated to the platinum. The article included a table of complexes depicting Pt(II) bonding with PR₃ ligands as Pt–NCS, but with NR₃, the bonding was Pt–SCN. No attempt was made by Turco and Pecile to prepare the linkage isomers. Since my family and I had already begun to pack for our return trip home, I decided to wait to talk to John (Burmeister) in person (**Fig. 3-13**). John had spent three years working on this problem, all without success. My first suggestion to John was that he read the Turco and Pecile paper, and then we would talk about what might be done.

We both agreed that if the M–NCS and M–SCN bonding on the same metal depends on the other innocent ligands coordinated to the metal, then it may be possible to find a borderline ligand that gives a mixture of the N and S bonding isomers. Burmeister had tried so many different approaches to the problem with no success that he was hesitant to try this because if it failed, he would have nothing but negative results in his dissertation. Nevertheless, he went ahead, deciding to use Pd(II), because its substitution reactions are much faster than the reactions of Pt(II). After investigating the effect of many ligands on the nature of bonding, it was decided to use $(C_6H_5)_3As$. We were finally rewarded by isolating the first metal linkage isomers of NCS! The following equations demonstrate the reactions (eqs. 3-19, 3-20).

$$[Pd(SCN)_4]^{2-} + 2(C_6H_5)_3As_2 \xrightarrow[C_2H_5OH]{0°} [Pd(As(C_6H_5)_3)_2(SCN)_2] + 2SCN^-$$
(orange)   (3-19)

$$[Pd(As(C_6H_5)_3)_2(SCN)_2] \xrightarrow[30\ min]{150°} [Pd(As(C_6H_5)_3)_2(NCS)_2]$$
(yellow)   (3-20)

One night during all this effort, I was awakened from a sound sleep at about midnight by the ringing of the telephone. On the other end was John, who was shouting, "I got it! I got it!" I asked, "What did you get?" and he said, "I got spectroscopic evidence of linkage isomers."

At long last, John was successful in obtaining the first example of the ligand NCS⁻ being coordinated to a metal having identical innocent ligands L with the bonding of $L_2Pd-(NCS)_2$ and $L_2Pd-(SCN)_2$. He was able to write a fine, acceptable, and positive dissertation, and bid us farewell with his Ph.D. Before leaving, John asked if, once he had an academic position, would it be OK if he continued his research on linkage isomers. I said, "Certainly, because we have other problems to work on." John had considerable success with his work on these systems, and he coined the word "schizophrenic" to describe an ambidentate ligand capable of coordination to a metal complex at two different ligand atoms. Burmeister published several papers on this subject. One was a review in *Coordination Chemistry Reviews* (1968) which received the *Current Contents* journal's "Week's Citation Classic" in 1988 for having been cited more frequently (270 times) than any paper published, to that point, in *Coord. Chem. Rev.*

Ligand Substitution Reactions of Pt(II) Square Planar Complexes

Early on, Pearson and I had most of our research group working on substitution reactions of octahedral complexes, but we certainly wanted to have a look at such reactions of four-coordinated complexes. We decided that the place to start was with the square planar complexes of Pt(II). One reason for this decision is because its reactions are slow enough to be followed by conventional means (since these were the days before instruments to investigate fast reactions were available). Another reason for our choice was the considerable amount of information available on the reactions of Pt(II) complexes. Much of the information was published by the Russians because Russia has good natural sources of platinum. The Russian government had established an *Institute of Platinum Chemistry*, where much of this research was being done. Of particular interest to us were the publications of I. I. Chernyaev on his discovery of the trans-kinetic effect and the publications of A. A. Grinberg, who was investigating the kinetics of Pt(II) complexes.

Our studies on the mechanisms of ligand substitution reactions of Pt(II) compounds were being done in the 1960s, a time when many persons in Eastern countries were not permitted to travel to the West. However, in 1962, on my way from Rome to Stockholm to attend an International Conference on Coordination Chemistry (ICCC), I was invited to stop in Krakow, Poland, to participate in a satellite meeting of about 20 coordination chemists. This meeting was arranged by two members of the Polish Academy of Science: Professor Boguslova Jezowska Trzebiatowska (**Fig. 3-14**), at the University of Krakow, who was an inorganic chemist, and her husband, a physical chemist. Since her name was so difficult to pronounce, we decided just to

Figure 3-14. Professor Boguslova Jezowska Trzebiatowska at the University of Krakow and member of the Polish Academy of Science. Because of the difficulty in pronouncing her name, we called her the "Madam T."

call her Madam T. The meeting was held at their house and, as I recall, three or four participants were from Russia. Chernyaev was one of the Russians that attended and, in his talk, he told us about his recent use of the *trans-effect* to prepare some new Pt(II) complexes. He was a delightful person and, through an interpreter, we had a very pleasant conversation.

I told Chernyaev how we had made good use of his trans effect, and that we were studying the kinetics and mechanisms of reactions of Pt(II) compounds. He said it was important that this be done, and that he would be reading our publications. It seemed he would have liked to talk about life in the US as compared to that in Russia, but he must have been at liberty to discuss only chemistry. Grinberg, who had published some significant papers on the kinetics and thermodynamics of Pt(II) complexes, was not present at this small informal gathering. I always hoped that I would meet him at some international symposium, but I never did.

Another person who had done some very elegant research on the synthesis and reactions of many Pt(II) complexes was J. Chatt. It was clear from his publications that he assumed that the reactions proceeded by a S_N2 process, so he was not at all surprised when we published a series of papers in keeping with this mechanism. In addition to his experimental work, which was always done with great care, he reported, independently, at about the same time as did M. J. S. Dewar, the nature of ethylene to platinum bonding in Zeise's salt, $K[PtCl_3C_2H_4]$ (Structure **15**, pg. 60). This salt had been prepared by the

Danish pharmacist Zeise in 1828 and its structure and bonding were not determined until the 1950s. This type of bonding has contributed immensely to the understanding of many of the systems in the very important class of compounds known as organometallics.

Chatt was a very fine, likeable person, but he was very British. For example, although our families knew each other and were guests at each other's homes, he was always called Joseph, never Joe. When he wrote to me, he always signed the letter Joseph. The first time he came to the US was in the early 1950s, when he was invited to give a talk at the Gordon Conference. I was his host and some of the attendees at the Conference came to me and said, "This guy acts as if he thinks he is superior to us. Tell him to 'knock it off'." I told them, "That is just the way British talk, try not to take it personally." By the end of the conference week, Joe had made many friends.

Our initial publication on the mechanism of ligand substitution of four-coordinate square planar Pt(II) complexes appeared in 1957. This study was made by Debabrata Banerjea, using NH_3/Cl complexes in aqueous solution. The rates of reaction with different nucleophiles were determined, and in each case it was found that the rates of reaction depended on the concentration and the nature of the nucleophile. This came as no surprise because of the low coordination number of Pt(II), which would readily permit attack at the metal with an increase in coordination number to five. In addition, Pt(II) complexes have a low energy, empty orbital that can easily accommodate a pair of electrons from the entering group. What *was* unexpected was that the reactions took place by a two term rate law. This suggests the mechanism shown by equations 3-21–3-23.

$$[Pt(NH_3)_3Cl]^+ \xrightarrow{\text{Slow } H_2O} [Pt(NH_3)_3H_2O]^{2+} + Cl^- \quad (3\text{-}21)$$
$$\xrightarrow{Y^-} \text{fast} \quad (3\text{-}22)$$
$$\xrightarrow[\text{fast}]{Y^-} [Pt(NH_3)_3Y]^+ + Cl^- \quad (3\text{-}23)$$

It was clear in the 1950s that there was a need for detailed kinetic studies of ligand substitution reactions of Pt(II) complexes, and our laboratory was prepared to do this because it was engaged in such studies of octahedral substitution. S. R. Martin, Jr., and his students, at about this time, initiated their investigations of such studies on aquation reactions of chloroammineplatinum (II) complexes.

There is now a large amount of kinetic data on substitution reactions of square planar complexes, all of which are explained in terms of a bimolecular (S_N2) displacement mechanism. For reactions such as

$$[MA_3X]^{n+} + Y^- \longrightarrow [MA_3Y]^{n+} + X^- \quad (3\text{-}24)$$

in water solution, a two-term rate law,

$$\text{Rate} = k_{H_2O}[MA_3X^{n+}] + k_y[MA_3X^{3+}][Y] \qquad (3\text{-}25)$$

is generally followed where k_s (solvent path, eq. 3-21) and k_y (reagent path, eq. 3-23) are first-order and second-order rate constants, respectively. Under pseudo-first-order conditions containing excess Y, the experimental first-order rate constant, k_{obs}, is related to the individual rate constants as shown by the equation

$$k_{obs} = k_s + k_y[Y]. \qquad (3\text{-}26)$$

This requires that a plot of k_{obs} versus [Y] be linear with an intercept of k_s for the reagent-independent path and a slope of k_y for the reagent path. Plots of this type are common for substitution reactions of square planar complexes. Such a plot is shown in **Fig. 3-15** for the reaction of *trans*-[Pt(py)$_2$Cl$_2$], with a variety of different reagents. The two-term rate law requires a two-path mechanism as that illustrated in **Fig. 3-16**, where S is the solvent and Y is the nucleophile.

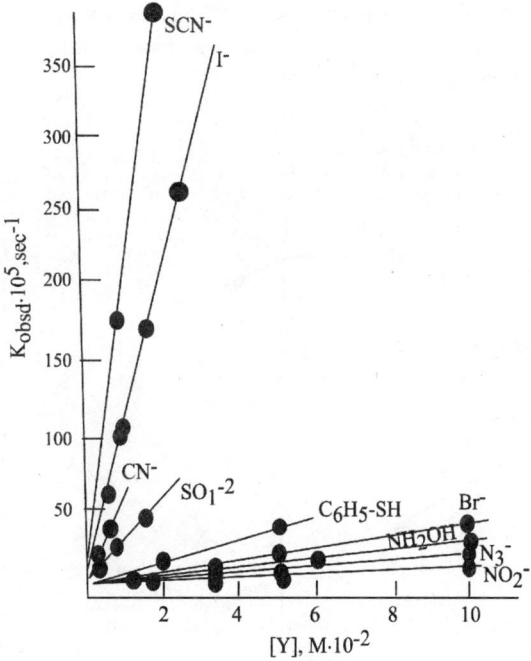

Figure 3-15. Rates of reaction of *trans*-PtCl$_2$(pyridine)$_2$ with different nucleophiles in CH$_3$OH solvent at 30°C.

The results shown in **Fig. 3-15** are consistent with Chatt's classification that a polarizable ligand atom such as S of SCN⁻ is a much better nucleophile towards a polarizable metal such as Pt(II) forming Pt–SCN. His classification refers to these metals as class B metals. Such metals do not interact strongly with a much less polarizable ligand. The other class of metals A are much less polarizable and interact more strongly with the much less polarizable ligand atoms as would be for Al–F. My former colleague Pearson has proposed the nomenclature "soft" for highly polarizable metals and ligand atoms, and "hard" for the less polarizable metals and ligands. Like Chatt, Pearson also states that "soft/soft" or "hard/hard" interactions are more stable than are "soft/hard" systems. Both Chatt and Pearson would be the first to agree that this classification is qualitative, albeit useful. I recall that the first time Pearson told me Chatt's A and B classes, as well as the work on the stabilities of metal complexes by Jannik Bjerrum and Gerold Schwarzenbach, my response was "So what else is new?" I was aware of Jöns Jacob Berzelius (1779-1848) who had discovered the elements selenium, silicon, thorium, and zirconium. I am told that he made the point that certain metals are found on the earth's crust as sulfides (HgS) and others as oxides (Al_2O_3). However, the names "hard" and "soft" have a real meaning, readily understood, and this acid/base concept is even in beginning chemistry text books, and in undergraduate courses in inorganic chemistry.

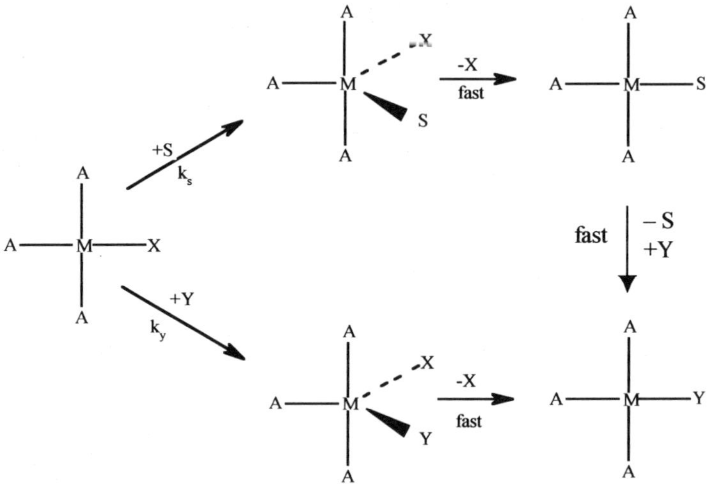

Figure 3-16. General S_N2 mechanisms of square-planar metal complexes, such as Pt(II) compounds, where S is solvent and Y is entering nucleophile.

Figure 3-17. Professors Umberto Belluco and Marino Nicolini at the University of Padova, Italy. Both were visiting scholars at NU and did research in my lab. Umberto has done research in coordination and organometallic chemistry. He was largely responsible for starting the journal *Inorganica Chimica Acta* and is its Editor-in-Chief. Marino's research was mostly in medicinal chemistry. He was the person who initiated the international conferences on the chemistry of anticancer metal complexes, and on the chemistry of technicium (Tc), used for imaging diagnosis of patients.

One study was done in collaboration with Professor J. Chatt, on compounds prepared by his student Bernard Shaw, and investigated by Pearson and our student Harry Gray. This paper was important, for at least three reasons. One was that it was among the first on the kinetics of substitution reactions of organometallic compounds of the type [(PtPR$_3$)$_2$(alkyl or aryl)Cl]. The new compounds were prepared by Shaw and the study of the kinetics of the replacement of Cl$^-$ by pyridine was carried out by Gray. The second reason that this study was important is because it was the first time that the relative rates of reaction of the nickel triad were reported. These were found to be approximately 5,000,000:1,000,000:1 for Ni(II):Pd(II):Pt(II), respectively. The third important result from this research involved the acquisition of quantitative data on the trans labilizing effects of various ligands in Pt(II) complexes. The results indicated that both polarization and the π-bonding theories of the *trans* effect are needed to account for our results.

Chatt and I had similar interests in chemistry, so we often attended the same scientific conferences. At one of these meetings, we talked about the above work that was in progress. He said "The other day Bernard was complaining that each time he sends Harry a new compound, the kinetic results arrive so quickly that I hardly have time to prepare the next compound." I was able to report to Chatt that Harry said he received the compounds so fast that he found it difficult to keep up with the compounds sent to him. Chatt and I had a good laugh and decided that if we managed to keep this going, in little time we would have enough information to publish.

One other aspect of work on Pt(II) was done in my laboratory by Umberto Belluco (**Fig. 3-17**), in collaboration with Lucio Cattalini and Professor Aldo Turco. This research was primarily proposed by Pearson. He had become interested in his concept of hard and soft acids and bases, and wanted to test this theory. It was clear that a good choice for such a test would be the reactions of Pt(II) compounds, since we knew their rates of reaction depended on the nature of the entering ligand. For example, although

OH$^-$ is a very strong base, it is a very weak nucleophile toward Pt(II) compounds, whereas thiourea is a very weak base but an extremely strong nucleophile attacking Pt(II) complexes. This is consistent with thiourea (or S ligand atoms, in general) being a "soft" ligand, interacting strongly with the soft metal Pt(II), while the latter interact poorly with the "hard" ligand OH$^-$. It was further shown that by using trans-[Pt(py)$_2$Cl$_2$] as a standard, a set of nucleophilic reactivity constants η_{Pt} can then be used to estimate the rate constant for any nucleophile toward a given Pt(II) complex.

My family and I had been at the University of Rome the year before Umberto and his family arrived to spend a year with us. I kept to the usual work culture of the US. However, the Italians went home for a large lunch and a sound sleep, returning at about 4:00 PM and working until 7 or 8:00 PM. This was the Italian way, and how Umberto worked at the University of Padova. At NU, he became so interested in his research that he very soon adopted the American work style. More than that, we often found him eating his lunch while collecting data on a Beckman-DU spectrophotometer in order to follow the rate of some reaction.

Another event that occurred during Umberto's stay always makes me wonder whether we may have had something to do with the advent of the journal *Inorganica Chimica Acta*, now an important journal in the field of inorganic chemistry. After a good dinner with wine, followed by "grappa," hosted at our house, we sat enjoying our after-dinner drinks. During our conversation, the subject got around to the Italian journal *Gazzeta Chimica Italiana*. Umberto said it was a terrible, useless journal that even most of the Italian chemists would not read. I said, "In that case, when you return, why don't you start a journal in inorganic chemistry, for it seems this field is growing rapidly and there may soon become a need for such a journal." I was not serious when I made this remark, but when Umberto returned to Italy, he talked to his mentor, Ugo Croatto, about this. Ugo, being a wealthy Italian, decided to go ahead and finance the journal. The two of them named the journal *Inorganica Chimica Acta*. Harry Gray and I were asked to be associate editors to help with US articles. The journal has been a big success. A few years ago it was sold to *Elsevier Science* with Belluco as Editor-in-Chief.

Organometallic Chemistry

My interest in metal carbonyls began before they were classified as organometallics, even before this name was coined to be used for compounds with at least one carbon metal bond. My good friend E. O. Fischer (Nobel Prize 1973) became editor of the new *Journal of Organometallic Chemistry*. At some conference at about that time, E. O. gave a talk during which he mentioned the word several times. During the discussion of his lecture I asked what he meant by organometallic, and he gave the above definition. I then made the point that the common complex [Fe(CN)$_6$]$^{3-}$ has six metal-carbon bonds but is called a metal complex or a Werner complex. He responded "Yes, that is good. We should continue to call such compounds, complexes." In fact, classifying compounds by this name has become so arbitrary that it is left to the authors to decide. A few years ago Professor Dietmer Seyferth, editor of the journal *Organo-*

metallics, wrote an editorial in which he concluded that, "An article can be considered an organometallic paper if it is of interest to organometallic chemists."

Now—back to how I became interested in metal carbonyls. My first sabbatical year was spent in Copenhagen as a guest of Jannik Bjerrum at the University of Copenhagen in 1954-55 (pp. 37-60). During a drive to Rome, my wife and I stopped in Amsterdam in order for me to attend the third ICCC. One of the plenary lectures was given by Professor Walter Hieber, (**Fig. 2-10**) who had spent all of his professional career investigating the chemistry of metal carbonyls. He always had 30 or more students and/or assistants working in his laboratory.

Figure 3-18. Professor Luigi Venanzi (1927-2000) at the Technical University of Switzerland (ETH). His excellent research dealt mainly on metal complexes with class B or soft ligands (see discussion in text).

As a result Hieber was so prolific in publications that he is often referred to as "the father of metal carbonyl chemistry," in spite of the fact that the first metal complex containing only CO ligands was $Ni(CO)_4$, discovered by Ludwig Mond in 1890. Mond made this discovery when he noticed that CO seemed to be causing corrosion of a nickel valve on a cylinder containing CO, another example of serendipity playing an important role in chemistry. The first metal carbonyl, $Pt(CO)_2)Cl_2$, was discovered in 1870 by P. Schülzenberger.

Hieber's plenary lecture was scheduled to last 45 minutes, but he talked for more than one hour. He did not know English and gave his talk in German. Fortunately, he had good slides, making it possible for those of us who knew no German to follow his lecture. After his lecture, I complimented him on his elegant work, but then I stated, "You never told us how some of these reactions take place—in other words, observations on the reaction mechanisms." Fortunately, Luigi Venanzi (**Fig. 3-18**) was able to act as our translator, and he told Hieber what I had said. Hieber looked directly at me and replied in German, "Young man (I was 35), we do real chemistry in my laboratory, not the philosophy of chemistry." This meant that they were doing reactions and syntheses, but cared less about mechanisms or theories of bonding. This was exactly what I wanted to hear, that there had been almost no work done on the kinetics and mechanisms of CO substitutions of metal carbonyls. It was easy to see how my research group could investigate the substitution reactions of these compounds in the same manner as we were doing on the Werner complexes. The metal carbonyls would allow us to investigate compounds having different coordination numbers and structures, such as tetrahedral $Ni(CO)_4$, trigonal bipyramid $Fe(CO)_5$, and octahedral $Cr(CO)_6$. At the time it was even believed that one might find a difference between terminal and bridging carbonyls in compounds such as $Co_2(CO)_8$. Since no such studies had been reported, this would result in original research suitable for publication and for students to use for their Ph.D. dissertations.

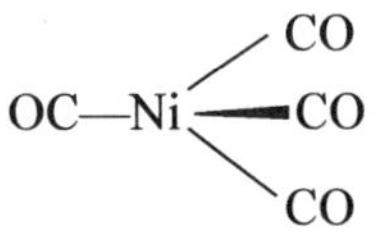

16 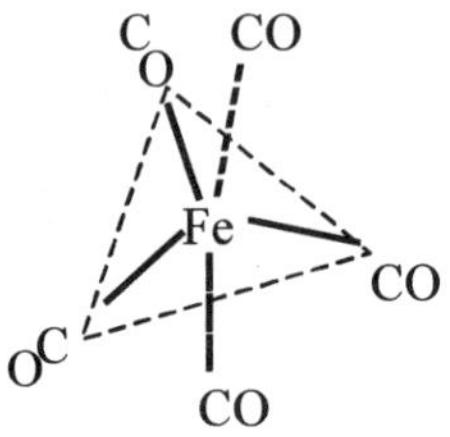   **17** 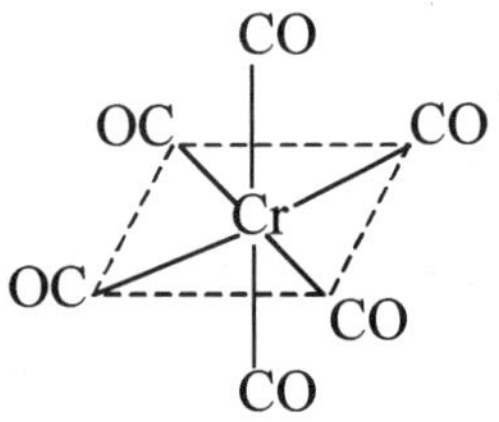   **18**

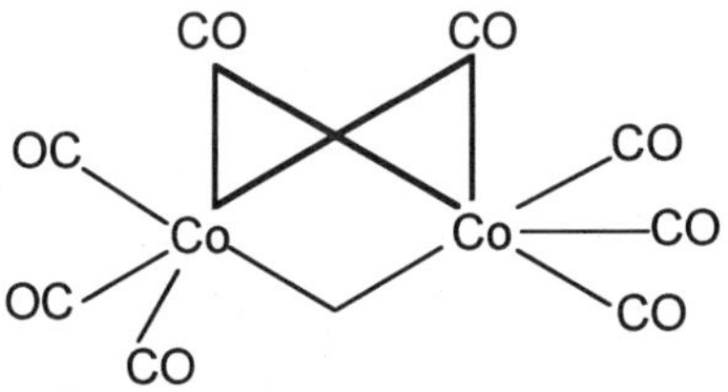

19

I couldn't wait to get back to NU and talk to beginning graduate students about this "wide-open" area of inorganic chemistry. I told students who wanted to do research in my group that we were in an ideal position to investigate the mechanisms of CO substitution, which would be similar to the work we were doing with ligand substitution of metal complexes. The only difference would be that metal carbonyls are toxic. Nonaqueous solvents would be required, and experiments would have to be done in the absence of air. (Just the opposite is true of metal complexes which are generally not toxic, and dissolve in water to give solutions stable in air.)

The students I talked to about this work were beginning the research for their Ph.D. dissertation. They would then go to the lab and ask the other graduate students about their research. In a couple of days, the student would come back to me, requesting that I suggest a research problem similar to that being done in our lab on Werner complexes. In spite of my enthusiasm and urgency for starting a research program on metal carbonyls, it began to appear that the new students felt more secure with a problem akin to those going on in our lab.

Fortunately a brave young son of Polish immigrants (as I am of Italian immigrants), Andy Wojcicki, decided to accept the challenge and initiate studies of CO substitution of metal carbonyls in our labs. He decided to start with the exchange of CO

Figure 3-19. Professors Andrew Wojcicki (Ohio State University) and Harry B. Gray (California Institute of Technology). Andy is to the left; I am in the center; and Harry is to the right. This photograph was taken in 1994 at NU, when Harry received the Basolo Medal, presented to him by Andy.

of the metal carbonyls with radioactive ^{14}CO. He constructed a set-up where CO was pumped through a solution of the metal carbonyl and the circulating gas was monitored for its radioactivity versus time using a Geiger-Muller tube. This cumbersome technique had to be used because it was before the day of ^{13}C nmr. What Andy found was that the relative rates of CO exchange for the simple molecular carbonyls varied in the order $Ni(CO)_4 \gg Cr(CO)_6 > Fe(CO)_5$— and that all of these react by a dissociative S_N1 mechanism. Even more important to me than this fine work by Andy was his enthusiasm and excitement, which meant that beginning students finally wanted to work on these systems. Excellent students at about that time who chose to do metal carbonyl research for their doctorate dissertations included Bob Angelici, Al Brault, Earl Thorsteinson, and Don Morris.

Before leaving the Wojcicki story, I want to mention a few things relating to Andy's graduate student years in our lab. To begin with, his lab mate was none other than Harry Gray (**Fig. 3-19**). Most chemists now know Harry, or at least know of him. One can imagine what it was like for Andy, working in the same lab as Harry. Andy was quite a serious graduate student, 100% devoted to his metal carbonyl research, whereas Harry was just the opposite, outgoing and interested not only in his work, but also in what others were doing in the group. These differences in personality is perhaps the reason the two got along so well, forming a close friendship, which is still strong today. Andy was not the athletic type, so he did not challenge Harry at tennis, but, as partners in bridge, they did manage to get a few master points.

The one thing generally well known among most of their professional brothers and sisters is how they coped with handling toxic metal carbonyls in a poorly ventilated lab. Since it is known that canaries have a lower tolerance for toxic fumes than do people, they bought a canary which they kept in a cage in the lab. They felt that they should be safe as long as the canary was alive. It was given the name Linus, and after a year or more, Linus died of old age. They then returned to the pet shop, but found it to be out of canaries. However, wanting to make a sale, the owner told them that a parakeet would serve the same purpose. The parakeet replaced the canary, and although it served to warn them of toxic fumes, it also took a peck at Harry when he was reaching in the cage to feed it. Harry came down with psittacosis, which caused him to have severe headaches each afternoon. At times the pain was so bad that he was not able to return to the lab in the afternoons. This condition finally subsided after a few years.

Andy and Harry provided yet one more unforgettable memory. Without Pearson's and my knowledge, they investigated the ligand exchange of $[Rh(PR_3)_2(CO)Cl]$ using ^{14}CO, and $^{36}Cl^-$. Bob Angelici provided the work on the PR_3 exchange. Andy and Harry wrote and submitted their paper to the *Journal of the Chemical Society Chemical Communications* without the knowledge of Pearson or myself, and without our names on the paper. The first Ralph and I knew about this was when we found a reprint of the article in our mail. The authors were Wojcicki and Gray, but not Angelici. A satisfactory explanation has yet to be given as to why they did not include Bob's name on the paper. Graduate students reading this should know that such behavior would not be tolerated by all mentors.

As mentioned earlier, Wojcicki opened the watershed on metal carbonyls in our lab and after that, other graduate students wanted to work on such problems. There were far too many studies to discuss here, so only one that evolved into a useful concept and quickly came to be used globally will be briefly told. This began with the research of Erlind Thorsteinson on $Co(CO)_3NO$ and, later, Donald Morris on $Fe(CO)_2(NO)_2$. These two nitrosyl carbonyls are isoelectronic with the stable 18e count and both are tetrahedral, as is $Ni(CO)_4$. The three compounds are as similar as different compounds can be. We anticipated that the CO substitution of the nitrosyl compounds would proceed by a dissociative S_N1 mechanism, as does $Ni(CO)_4$. It came as a complete surprise to us when we found that the nitrosyls react by an associative S_N2 mechanism. We concluded that the NO group must be behaving in some different way than CO. Since an S_N2 reaction requires attack on the metal by a nucleophile, it means that the electron count must go from the stable 18e to the unstable 20e. We began to think of how the presence of NO might prevent the transition state or active intermediate from having 20 valence electrons. It was thought that this could only happen if a pair of electrons, normally on the metal, were localized on the nitrosyl ligand. This seemed reasonable, because N is more electronegative than is C. This meant we had to localize a pair of electrons on N in order to provide an empty low energy orbital on the metal to accommodate the entering electron pair of the nucleophile. This was done as shown in equation (eq. 3-27). We then made use of a bent sp^2 M-$\overline{N}\diagdown_O\diagdown$ which had never been

$$\text{Co(CO)}_3\text{NO} \xrightarrow[\text{slow}]{+L} \text{(CO)}_3\text{CoL} \cdots \overset{\cdot\cdot}{\underset{O}{N}} \xrightarrow[\text{fast}]{-CO} \text{Co(CO)}_2\text{NOL} \qquad (3\text{-}27)$$

18-electron   18-electron   18-electron

reported. After discussing this in BIP, my colleague, Professor James Ibers and his students, decided to attempt to isolate a single crystal of a stable, bent M−N−O. This was done by adding NO⁺ to the Vaska compound IrCO(PR$_3$)$_2$Cl. Their x-ray study, done by Jim, was the first to show a metal−NO bent structure. Now, several examples are known to have this structure.

This phenomenon of localizing a pair of electrons on a ligand, thus permitting an S$_N$2 substitution, was news to us and, we thought, to other coordination chemists. Therefore, we decided to look for other ligands that would serve the same role as NO. Fortunately, very soon we found that cyclopentadiene could behave as such a ligand. At the time, we had an arrangement with E. O. Fischer to exchange predoctoral students during the last year of their dissertation research. This was the idea of Oren Williams of NSF, and was to be a trial program, but it was never repeated because of some NSF policy. My students would go to Germany and work in Fischer's lab to get experience with the synthesis and handling of organometallic compounds. His students would come to my lab and learn about investigations of kinetics and mechanisms of such compounds. The first student to arrive from Fischer's lab was Hans Schuster-Woldan. He had just completed his initial dissertation research on the synthesis of the cobalt triad of η^5-C$_5$H$_5$M(CO)$_2$ compounds, where η^5-C$_5$H$_5$ indicates that cyclopentadienyl (Cp) is bonded to the metal equally by the five carbons of Cp. Thus, the term, η^x, indicates the number of atoms of a compound bonded equally to the metal, for example, η^3-Cp tells us that only 3 of the 5 carbon atoms of Cp are bound to the metal.*

Hans discussed with me his research in Germany, where he prepared and studied the reactions of (η^5-Cp)Rh(CO)$_2$ with phosphines. He said the substitution reaction, shown by equation (3-28), proceeded at a moderate speed that could be followed easily by conventional means. He suggested that we study the kinetics and mechanism of this reaction. It was not long until he had completed a couple of kinetic runs, and we were able to see that the reaction did indeed proceed via an S$_N$2 mechanism.

$$(\eta^5\text{--C}_5\text{H}_5)\text{M(CO)}_2 + L \rightarrow (\eta^5\text{--C}_5\text{H}_5)\text{M(CO)L} + CO \qquad (3\text{-}28)$$

M = Co, Rh, Ir; L = PR$_3$ or P(OR)$_3$

---

*The use of η was introduced by F. A. Cotton (**Fig. 5-13**) to indicate the number of equivalent atoms bonding to a metal—here the first structure has 5 equivalent carbons bound to Rh thus η^5, while the second structure has three bound, thus η^3.

As a real pedestrian with chemical theory, only able to count valence electrons, I viewed η^5-Cp as a 6e donor analogous to 3COs. This would mean that $(\eta^5$-Cp)M(CO)$_2$ can be thought of as being pseudoisoelectronic with Fe(CO)$_5$. However, our earlier study showed Fe(CO)$_5$ to be extremely slow to undergo CO substitution by an S_N1 process. This then seemed to mean that, like the nitrosyl carbonyl, there must be a ligand present that permits the localization of a pair of electrons on it. This had to be the η^5-Cp group, so we suggested formation of an active intermediate by localizing a pair of electrons on Cp, giving a η^4-Cp with the fifth carbon being an anion. We were wrong in our suggestion for the transition state, but correct in having to localize an electron pair on Cp. Ten years later, Dr. Robert Cramer (1913-93) of the Du Pont Chemical Company said the intermediate may have a η^3-allyene-ene bonding of the Cp group, and a year after that, Professor H. H. Brintzinger reported the structure of a η^3-Cp metal compound. (See eq. 3-29 for the structural representation of this reaction.)

$$\underset{\substack{\text{Rh(CO)}_2 \\ \eta^5 \text{ 18-electrons}}}{\text{[Cp]}} \xrightarrow[k_2,\text{ slow}]{+L:} \underset{\substack{\text{LRh(CO)}_2 \\ \eta^3 \text{ 18-electrons}}}{\text{[Cp]}} \xrightarrow[\text{fast}]{-CO} \underset{\substack{\text{LRhCO} \\ \eta^5 \text{ 18-electrons}}}{\text{[Cp]}} \qquad (3\text{-}29)$$

Our research continued to look for other groups that might cause ligand substitution by this type of mechanism. Following is our list of such groups: NO, Cp, carboranes, SO$_2$, and indenyl (Id). The reactions of Cp metal compounds which proceed by an electron pair localizing on the Cp and causing the change of bonding from η^5-Cp to η^3-Cp to η^5-Cp (see eq. 3-29) is referred to as *ring slippage*.

There had been a report in the literature that CO exchange of CH$_3$M$_3$(CO)$_5$ takes place by a S_N2 mechanism. We were very interested in this because our experience was that six-coordinated complexes react by a S_N1 process. Our postdoctorate Roger Mawby investigated the reaction in eq. 3-30 and found that it reached a concentration of the entering ligand L where it no longer depends on its concentration, thus it is a S_N1 process. We suggested a methyl migration to an adjacent CO to form the acetyl group (eq. 3-30). Later this mechanism of one ligand moving onto a nearby one came to be called a *ligand migration reaction* and has had an important role in organometallic chemistry. When Roger returned to a faculty position at York University in England, he and his students continued to study this reaction. One important thing they found is that the indenyl compound reacts 10^3 to 10^4 times faster than the analogous cyclopentadienyl compound.

$$\text{(3-30)}$$

It was not clear to us why this rate enhancement of Id over Cp compounds should occur. In spite of this, we believed it would be worthwhile to see if a similar effect would be observed for CO substitution as was found for $(\eta^5\text{-Cp})Rh(CO)_2$. Mark Rerek and Liang-Nian Ji prepared the $(\eta^5\text{-Id})Rh(CO)_2$ and determined its rates of reaction with various phosphines (PR_3). They found that the rates of reaction (eq. 3-31) depend on the

$$\text{(3-31)}$$

Rh(CO)$_2$ LRh(CO)$_2$ LRhCO

η^5 18-electrons η^3 18-electrons η^5 18-electrons

nature and the concentration of the phosphine, as it did for the corresponding Cp compound, and both involve an S_N2 mechanism. The difference between the two was that reactions of the Id compounds were roughly 10^8 times faster than those of the Cp compounds. One reason for this may be that shown by eq. 3-30. Here we see that the starting compound has a six-membered ring fused to Cp, and x-ray determined structures show that two of the carbon-carbon bonds in the six-membered ring are shorter than the other C–C bonds. This means that the shorter bonds indicate double C=C bonding as shown in eq. 3-31.

However, in the mechanism we proposed for ring slippage, the active intermediate results in formation of a fused benzene ring. It then appeared that the driving force for an Id compound over an analogous Cp compound is the formation of a very stable benzene ring. Since the rate enhancement was so large, we decided to name it the *indenyl kinetic effect*. It is a pleasure to know that this term and this discovery has been widely used globally.

One other example of grossly exchanged rates of reaction of analogous organometallic compounds was examined in detail in our lab. This was prompted by the fine research of Professor Ted Brown at the UI and his students. They studied the photochemical reaction of $(OC)_5Mn\text{-}Mn(CO)_5$, obtaining the 17e species $Mn(CO)_5$ as an active intermediate, which was observed to very rapidly undergo CO substitution. Since I was aware of the earlier work done by Fausto Calderazzo (**Fig. 3-20**) and his colleagues on $V(CO)_6$, we decided to investigate this well-defined, stable 17e compound. Professor Bill Trogler (now at the University of San Diego, California, **Fig. 3-21**), Professor Qi-Zhen Shi (now at the Northwest University in Xian, PRC), and Tom Richmond found that CO substitution in $V(CO)_6$ is very fast, proceeding via a S_N2 mechanism. From our work and that of others, it was known that CO substitution of the corresponding stable 18e $Cr(CO)_6$ is slow and involves a two term rate law. One path is dissociative S_N1 and the other path for reactions with strong nucleophiles is associative S_N2. Using n-butyl phosphine as the entering ligand, at room temperature, it was found that the associative pathway of 17e $V(CO)_6$ reacts 10^{10} times faster than does the stable 18e $Cr(CO)_6$. However, the 18e anion $[V(CO)_6]^-$ is substitution inert and only photochemical replacement of CO has been reported. These results afforded the first quantitative comparison of rates of ligand substitution between analogous 17e and 18e compounds. Professor Qi-Zhen Shi (**Fig. 3-22**), who worked on this research, was the first Chinese scholar to come to do research in my lab. My first postdoctorate in 1953 was Yun-ti Chen from Nankai University. Later, I will discuss how this all resulted in my seven visits to the PRC (see *Countries and Chemists Visited*).

Figure 3-20. Professor Fausto Calderazzo of the University of Pisa is best known by inorganic chemists for the diversity of his research. His work starts with the 17e $V(CO)_6$ followed by other organometallic chemistry, and synthetic O_2 carriers.

Figure 3-21. Professor William C. Trogler at the University of California, San Diego. When he was my colleague at NU, we collaborated on the mechanisms of reactions of 17e metal complexes. He is now doing research on the environment. Bill here is at Mendeleev's desk in the Mendeleev Museum, St. Petersburg, Russia.

Figure 3-22. Professor Qi-Zhen Shi and me. He was a visiting scholar in my lab for a year. We collaborated on research for a few years when he was at Lanzhou University. He is now at Northwest University in Xian.

Synthetic Oxygen Carriers

Since NIH deals with the important area of health, Congress always appropriates it more funds for research than it appropriates to other governmental agencies. I did not know of any inorganic chemists being supported by NIH, but I decided to send them a proposal on coordination chemistry of biological interest. My proposal pointed out that we were involved in basic research on the reactions of small, well-defined metal complexes in aqueous solutions. The proposal stated clearly that the difference between these complexes and the much more complicated biological systems is that the latter have large organic molecules attached to the metal. However, when the focus is on the metal, one finds that the number of ligand atoms attached to the metal is the same, as is the stereochemical distribution of the atoms on the metal. I suggested that biochemists and molecular biologists might be able to make good use of the information obtained in our studies. Certainly this would surely be the case for their research on metalloenzymes and metalloproteins. This was enough for NIH to fund the research.

Fortunately, at the same time that we were doing this research supported by NIH, I was asked to serve on one of their research proposal panels. At my second meeting on the panel, we were told that Congress had decided to provide government agencies funding for scientific research only if the research was specifically related to the mission of the agency. This meant that NIH- supported research would have to be specifically related to some health problem. It was almost certain that the research we were doing would no longer pass this test and receive funding. Therefore, I had to think of some other problem. This was not too difficult for I had wanted one member of our research group to have a look at the chemistry of synthetic oxygen (O_2) carriers of cobalt. I had read the papers published by Melvin Calvin (1911-97; Nobel Prize 1961) after World

War II.* Their work involved the interaction of a solid cobalt complex with gaseous air. What I proposed we do was investigate the solution kinetics and mechanism of O_2 uptake and release by the most promising cobalt complex reported by Calvin. His research group used primarily different salen cobalt complexes (see **20**). They found that when one attached group was fluorine (F) on each of the benzenes, the compound was the most effective and stable. In all the cases of oxygen uptake of these complexes, it was necessary to have a ligand L in the axial position (see **22**).

20, Co(salen)

My graduate student Al Crumbliss decided to work on this problem, and he started with a library search of how to prepare the desired salen compounds. He found that it would be easy to attain high yields of the compounds because they are insoluble in almost any solvent, so the compounds readily separate in pure form from the reaction mixtures. Al also realized that salen would not be suitable for our needs, since we wanted to study the oxygen carriers in solutions of different solvents. However, he did find that the analogous compound acacen (see **VI**) was reported to be soluble in various

21, Co(acacen)

---

*The research had been done during the war with funding from the Navy. If successful, the cobalt complexes would be used by submarines, allowing them to surface and collect O_2, then release it as needed when submerged. Since the research was classified, it could only be published after the war. Years later, I was asked by the U.S. Air Force to be a consultant. The problem was that the commercial Co(salen) they purchased varied from batch to batch. After some time, this problem was solved. This synthetic oxygen carrier was to be used in the B-1 bomber instead of cylinders of liquid oxygen. This meant that a direct hit of fire from the enemy would not cause the disaster posed by any rupture of the oxygen cylinders. It would also mean that the bomber would not have to stop somewhere to take on cylinders of oxygen. The engineers were able to install the Co(salen), and have it function properly. All of this met the specifications of the Air Force, but then the government did not appropriate the funds needed to build the B-1 bomber.

solvents. This compound would satisfy our needs, providing it behaved as an oxygen carrier, which it did. The experiments designed to investigate the rates of O_2 uptake failed, because there was instant formation of the O_2 adduct. Since we would not be able to measure such fast reactions, it was decided to measure the extent of O_2 uptake of the compound by monitoring changes of the ligand attached to Co in the axial position. Back as far as the time of Alfred Werner, it was known that if Co complexes were prepared in the presence of O_2, the products are bridged dimers such as $[(H_3N)_5Co-O-O-Co(NH_3)_5]^{4+}$. These compounds have been extensively studied, and their structures have been determined by x-rays of single crystals.

Independent of our work was that of Professors Carlo Floriani and Fausto Calderazzo that appeared in print at about the same time (1969) as did ours. They investigated the O_2 adducts of several Co salen complexes and were able to isolate one as a solid, which contained one Co to one O_2. Al and I were surprised to see that Al's addition of O_2 to a Co(acace)L with different axial ligands L resulted in the amount of O_2 added reaching a plateau of one O_2 to one Co. We were expecting this uptake of O_2 to level off at one O_2 to 2Co's, forming the usual bridged type compound (shown above). That these species in solution were 1Co to 1O_2 complexes got us very excited because, if true, then here was a synthetic O_2 carrier that resembled hemoglobin (Hb) and oxyhemoglobin (HbO_2). I had long known of the bridged Co compounds and of solution kinetic data to form 1Co to 1O_2, but never had such compounds been isolated as solids and characterized.

Co(acacen)L, 22   23

For this reason I kept after Al to get as much information as possible on his crystalline products: elemental analysis, molecular weight, amount of O_2 released by the solid O_2 adduct, and various spectroscopic measurements, including epr (electron paramagnetic resonance) done and interpreted by my colleague, Professor Brian Hoffman (**Fig. 3-23**). These results suggested that O_2 is bound as a superoxide to Co, making the formulation $[L_5Co(III)O_2^-]^{3+}$. The epr measurements showed that the electron is localized on the O_2, and the infrared spectrum has a stretching O–O band in the region of a superoxide, such as KO_2. The epr results also supported an end-on-bent Co–O_2 (see **23**). What was still needed was a structure determined by x-ray of a single crystal. My colleague, Jim Ibers, was prepared to do this for us, providing we gave him a suitable single crystal. For one month or more, Damon Diemente tried to grow such a crystal, but with no success. Shortly after our attempts failed, Professor Ward Robinson of Canterbury University, New Zealand, reported the x-ray structure as containing a bent Co-O_2. This is also the structure of Fe-O_2 in oxyhemoglobin (Hb-O_2)$_2$. Robinson later told us that one of his undergraduate students readily grew the crystal. Damon was certain that this must have had something to do with their being "Down Under." In any case, their x-ray structure showed that our proposed bent structure was correct. At one of our informal research group meetings, the importance of the axial ligand on the O_2 affinity of Co(acac)L was the topic of discussion. At some point, my postdoctorate, David Petering, suggested that we should try globin. My response was, "What is globin?" Dave proceeded to explain to me and my research group that it is the globin part of hemoglobin. Our response was "Yes, it should work, because all other axial ligands that we have tried have worked." Dave tried, and it did work. He had gotten his Ph.D. at Michigan University as a physical biochemist and was comfortable working with biological systems, but I was not, so I suggested that he should work on the problem with Hoffman. He did, and they gave birth to *coboglobin*. This was Brian's start with research on such systems, and the rest is history.

Also, very soon after our paper had appeared in print, I had a letter from A. Veillard telling me they had read our article and had conducted an *ab initio* calculation on the molecule, which also showed it should have a bent Co-O-O configuration. I was glad to have them write and tell me this, but, as I do not understand such theoretical work, I just had to be pleased that they had so quickly read our paper and responded to it.

It is of interest to note that we had a similar response to our paper reporting the discovery of a synthetic O_2 carrier of manganese. My colleague Brian Hoffman obtained the epr spectrum of the O_2 adduct and concluded that it was $[L_4Mn(IV)O_2]^{2+}$, where O_2, as a peroxide, occupies only one coordination site and has a T structure (see **24**). This was likewise supported by the O_2 stretching frequency observed in the infrared spectrum. Again, I had a letter and preprint from the chemists in Strasbourg, but this time their calculations did not give the structure we proposed. However, this did not trouble me for I could not understand their calculations and, in any event, would always give preference to experimental results over theory. Some months later we were pleased to see an article by Professor Michael Hall at the Texas A&M University

Figure 3-23. Professor Brian Hoffman of Northwestern University with one of his four daughters (Julia) at the wedding of another (Tara).

Faculty Position at Northwestern University

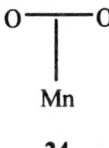

24

in which he found an error in the previous calculations. Making the necessary changes showed that our proposed structure was in keeping with theory after all.

As mentioned earlier, we had to do research on problems directly related to the mission of the governmental agency in order to get NIH funding. Our work on synthetic O_2 carriers was supported by the Heart and Lung Institute of NIH. Since some congressmen are known to die of heart attacks or lung cancer, perhaps this Institute gets more than its share of the money appropriated to NIH by Congress. We were pleased about this, for it provided continuous support of our research.

The importance of research in this area was made apparent when one day I received an invitation to come to the NIH and attend a conference on artificial blood. My first reaction was to tell them that I knew nothing about real blood, and even less about artificial blood. Then I realized I had to go because they were the source of the support for our research. When I checked in at the hotel, I was given a copy of the two day conference and a list of names and address. I looked over the names of roughly 100 attendees and recognized only two names, mine and that of Professor Jack Baldwin, then at MIT, now at Oxford. We were later to learn that we were the only two chemists. The other attendees were mostly medical doctors or pharmacists, whom I did not know. The research they were doing involved almost exclusively fluorocarbons which are good solvents for O_2. Most of us recall seeing a white mouse submerged in a colorless liquid (fluorocarbon) that survived because of the O_2 in solution. We heard about the research of many investigators, some from other countries. Most of them showed us pictures of different body organs that demonstrated the horrible side effects of their use of fluorocarbons. They used commercial fluorocarbons and often found different batches gave different results. As chemists, Jack and I were asked about this. We told them that the commercial products they were using were surely not pure. They needed to use purified samples of fluorocarbons if they were ever to expect reproducible results.

There was one successful result reported by a medical doctor from Harvard. He showed very descriptive and interesting slides. The first slide showed how blood was removed, while, at the same time, the milk white emulsion of a fluorocarbon was introduced. The next slide was a close-up photo of the nose and ears of the white mouse. As we know, a normal white mouse has a pink nose and ears, but this mouse had a white nose and ears. We were told the "artificial blood" kept the mouse alive for a few days as its body slowly replaced it with real blood. The experiment had been done a year earlier, and the mouse was alive and perfectly normal, including its sexual activity. It is this type of result on people that would be extremely important. The patient would only use small quantities of the artificial blood as needed so that the body could more quickly excrete it and replace it with its own blood. Some successful cases have been reported with people.

Baldwin reported on his "capped" porphyrins (see **25**) where A represents CH_2 groups, the number of which allow the cap size to be changed. These capped porphyrins are synthetic O_2 carriers. There was much interest in these compounds for possible use as a synthetic blood. Jack was able to tell them several reasons why it would not work. He was also able to tell them about Jim Collman's picket fence porphyrin and why it would not be suitable.

$A = n\text{-}CH$

25

I had another interest in Baldwin's capped porphyrins. We were interested in using these to measure the oxygen uptake of iron porphyrin to give the adduct (porphyrin)LFe–O_2. Jack and I began a collaboration, which lasted a few years and resulted in four joint publications. We measured the uptake of O_2, CO, and NO, all on the same capped (porphyrin)LFe–X. The result showed the relative affinity of X to increase in the following order, $O_2 <$ CO $\ll$ NO. It was further observed that the affinity of O_2 and of NO decreased in relatively the same amount with an increase in the size of the cap, whereas that of CO remained about the same. We attributed this to the fact that O_2 and NO form bent FeX structures, **26**, while that of Fe–CO is linear, **27**. We attributed this to the peripheral steric effect of the larger number of methylene groups. Instead, the cap causes little steric effect on the linear structure of Fe–CO.

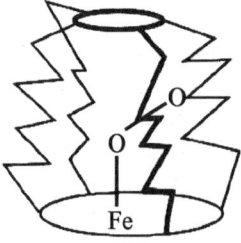 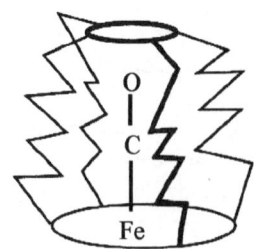

26, peripheral effect   **27**, central effect

* * *

Figure 3-24. Graduate students and postdoctorates of Pearson's and my research groups in the late 1950s-early 60s.

These were some of the areas of research that my students, postdoctorates, collaborators, and I visited during my 44 years of teaching and 47 years of research at NU. Clearly, I was not able to mention the research of all my students and postdoctorates. Their work, too, was of high quality and quantity, and I am more than pleased, and feel extremely fortunate, to have had the opportunity to work with all of them. The names of all of my students, postdoctorates, and senior collaborators are listed in the Appendix.

CHAIRMAN OF THE CHEMISTRY DEPARTMENT

Since 1952, our chemistry department operated with a rotating chairman who served for a period of five years. When the Dean of the College of Arts and Sciences asked me to become departmental chairman, I agreed to do so, providing that two conditions could be met: (1) my appointment would be for only three years, because I did not want to take five years "away" from my teaching and research; and (2) the department would hire an assistant professor in inorganic chemistry because we would need help with teaching and research mentoring for students of inorganic chemistry. The Dean met both of these requirements and I became chairman of our department for the three years 1969-72.

One of the things I wanted to do as chairman was to arrange an Industrial Associates Program for our department. This was prompted as a result of having dinner one evening with Henry Taube. He told me that their chemistry department at Stanford University had an Industrial Affiliates Program which was the best thing that the department could have done. Not only were funds from fees paid by affiliate members helpful, but even more important was the educational value of having industrial chemists spend two days a year in their department. Henry was so enthusiastic about their program that I believed our department should also have one.

During my chairmanship, I also was a member of the US National Committee for the IUPAC. IUPAC depends on funds obtained from member countries, such funds based on the country's GNP from chemical industries. As a member of this committee, I had to contact CEOs, Presidents, and/or Research Directors of US chemical industries and remind them of their obligation to support the IUPAC. This meant that I had the names and addresses of people whom we could invite for two days, as our departmental guests, to become acquainted with our department and students. We sent out 70 invitations, and had about 30 invitees attend. The two day program (5/21-22/1970) started with a welcome by President Strotz, and ended in an informal discussion with a panel of guests, faculty, and students. Unfortunately, this gathering was held during a difficult period—the Vietnam War (see p. 207) and a period of recession. After some discussion, it was decided that the time was not right to start an Industrial Associates because if it failed, it would make it more difficult to start one later.

However, we did begin the Charles D. Hurd Lectures in the academic year 1970-71. These yearly lectures continue and are given by an industrial chemist who spends three days with us. Our first lecturer was Wesley T. Hanson of Eastman Kodak, where I was a consultant at the time. The lecturers are asked to give one of their three lectures on company affairs and management. They are also given ample time to interact with students, postdoctorates, and faculty, giving all of us a chance to learn about chemical industries. The lectures provide an educational experience we would not get otherwise.

Fortunately, largely due to the efforts of Louis Allred during his service as departmental chairman, we now have an Industrial Associates Program, instituted on 01 January 1981. One or more chemists of industries that are members of the Associates attend a 2½ day conference which is directed toward a specific topic, such as catalysis, bioinorganic chemistry, organic syntheses, etc. Talks are given by the industrial chemists and members of our faculty. Students and postdoctorates hold poster sessions and discuss their research with our industrial guests which, at times, results in an offer of employment by some member of our Research Associates. Like the Charles D. Hurd Lectures, this contributes to our otherwise usually academia-based education.

As mentioned, my chairmanship came at the time of the Vietnam War when there was student disturbance on university campuses. One midnight I was awakened by the chairman of our geology department asking me to come to a meeting that students were holding in an attempt to enkindle the conviction that secret military research was being done in Tech. They proposed the destruction of lab apparatus in order to prevent such research. We tried unsuccessfully to inform them that none of the research was secret since it was all published in the open scientific literature. However, the students' suspicions were enough to cause our chemistry faculty and staff to stand guard 24 hours a day in each of the labs. Fortunately, all that the students accomplished was to prevent us from entering Tech for two days, except to take care of items needing daily attention.

The recession, also occurring during my chairmanship, caused industry to downsize and they let a number of research chemists go, even many who had been employed for several years. Some of these chemists were our alumni, and I had many telephone calls asking if our department could help them find employment. We were successful at times, but many had to wait until the end of the recession period when the job market for research chemists improved.

After two stressful years as chairman, I knew I did not want to continue beyond the agreed-to three years. At the beginning of my third year, I called the Dean to remind him that this was my third and last year, so he should plan on appointing my replacement. His response was "Fred, this is the earliest you have been on any deadline, so I know you must mean it." The departmental chairmanship continues with this three-year rotation, with a few exceptions when faculty have served a second three-year term.

4

OTHER ACTIVITIES

NATIONAL ACADEMY OF SCIENCES (NAS)

I was elected a member of the NAS in 1979, but before addressing my activities in the Academy, I want to mention how I first learned about my election. In 1979 I received a Fellowship from the Japanese Society for the Promotion of Science (JSPS), financed by Japan which required me to spend a month there giving lectures at several universities. I had been to Japan before and I liked the Japanese and enjoyed their food and saki. Here are a few things that I recall about that month in Japan.

I knew that the book *Coordination Chemistry* for beginning students by Ron Johnson and myself had been translated into Japanese and was being used in their freshman classes. What I did not know was that, at most of the universities where I gave talks, the students would line up with copies of the book for me to autograph. This pleased me, for it indicated that they liked the book and were making good use of it.

I also cannot forget the devastation left by a recent earthquake that had occurred in the Sandia region. There were large cracks in parts of the highway, and considerable damage was sustained at the chemistry building of Tohoki University. My friend and host, Professor Kazuo Saito, (**Fig. 4-1**) showed me their lab. It had been totally burned when solvent containers, broken during the earthquake, caused the flammable solvents to ignite. Large, heavy, and expensive instruments had been tossed around so severely that the equipment was beyond repair. Needless to say, when the lab is restored it will have the solvent containers locked in a strong metal safe, and the large instruments will be chained to the floor. After visiting the labs, a picnic had been arranged since it was a nice sunny day, so I had a chance to interact with the students.

My lecture schedule allowed some time for other activities. One evening in Hiroshima, I went to see a baseball game. The game is the same as ours, except there are no hot dogs—only Japanese food. The Japanese teams are allowed to have two foreign members on their teams, and most of these are US players who could not make the majors here. However, some of the Japanese are also very good—two Japanese pitchers are now playing in the US.

It is generally known that, at times, important business deals are initiated on a golf course. Here is an example of how it happened to me. At Osaka University, my friend Professor Otsuka who consults for the large chemical company, Ube Industries (considered the duPont of Japan), arranged for me to spend a day there— giving a talk and discussing some of their research with them. Otsuka and I had played golf twice together when he was visiting in the United States. Our games were about the same, none too good. Ube Industries has a beautiful guest house surrounded by *two* 18-hole golf courses (since Japan has a shortage of land, even one course is a luxury) which are used to indulge their golfing customers. Otsuka had arranged for us to arrive at noon the day before I was scheduled to visit Ube, and to leave at noon the following day. This allowed us to play golf twice, and so a foursome was arranged. One of our foursome was the executive vice president Sadao Tanimoto and the other was his deputy. We all played about the same, and it was interesting that we all had women caddies.

The four of us also played the following day, before we left at noon. During this round, Sadao asked me if I would be willing to have one of their research chemists come and work in my lab. I told him that I had the space and facilities and, when I had the funds to support him, would like to have him come. Sadao said that this would be no problem, since they would take care of the chemist and his family, and pay for all his expenses in the lab. As a result of these two golf games, I received four years of outstanding research without using any of my funding.

Toshiaki Hashimoto and his family were the first to come to Evanston and he spent two years working in my lab. During this time, he, his wife, and their two children became accustomed to US culture and their English improved. Later, he was appointed

Figure 4-1. Professor Kazuo Saito (1923-98), of the University of Tohoki in Japan. After retirement, he moved to the Institute of Molecular Science, where he was director of research centered on coordination chemistry. He was very active in the International Conference on Coordination Chemistry (ICCC), and we often met at these conferences.

Other Activities

to Ube's business office on Fifth Avenue in New York City for ten years. In my lab, he was succeeded by Makoto Shimizu who also spent two years. This clearly testifies that playing golf can have its benefits.

Now, as to where I was when, and how, I learned that I had been elected to the NAS. When I returned from the Ube Industries visit, my very good friend Professor Shoichiro Yamada (**Fig. 4-2**) was waiting for me at Osaka University. He said that a telegram had arrived for me, and I immediately thought that something serious had gone wrong at home. Instead, the wire was from my dear wife Mary, telling me the good news that I was elected a member of the NAS. She said that she and the children were all excited because they knew how important this was for me professionally. She also wired that she was receiving many telephone calls from colleagues, students, and others who wanted to congratulate me. I also received many telegrams in Japan which were sent to Professor Shigeo Kida at Kyushu University, my host JSPS University.

This honor bestowed upon me must be the pinnacle of my career. Being a member of the NAS is regarded as the highest honor that a US scientist can achieve, and the same is true for scientists of other countries in their National Academy of Sciences. Currently, the NAS of the US has a total of 2,255 members, encompassing six different areas of science. Chemistry is in the area of Physical and Mathematical Sciences which consists of mathematics, astronomy, physics, chemistry, geology, and geophysics. Chemistry now has a total of 204 members and, on an average, elects four new members each year. The total number remains about the same, because the number of deaths each year is about the same as the number added. The number of ACS members is 165,000, with several thousand chemists who are not members. The 204 NAS members are therefore a very select group of chemists. In my opinion there are as many, or more, well qualified chemists who should be, but never become, members of the NAS.

The NAS is a private, not for profit society of distinguished scholars engaged in scientific research, dedicated to the furtherance of science and technology and to their use for the general welfare of mankind. Upon the authority of the charter granted to it by Congress in 1863, the NAS has a mandate requiring it to advise the Federal Government on scientific and technical matters.

Figure 4-2. Professor Shoichiro Yamada, of the University of Osaka in Japan. He was in the teaching college of the University but managed to find time to do some research. He translated our book *Coordination Chemistry*, which immediately became adopted by most of the Japanese schools.

I have been reasonably active in the NAS during my 21 years as a member. I have attended all of its annual meetings since my membership, except for two times due to health problems. The meetings are always held at the NAS headquarters in Washington, DC. The meetings last four days, with all receptions and meals served at the Academy. This gives the members considerable time to meet and talk with one another. Much of my time has been spent talking to members about strong candidates for membership. Naturally, I am chiefly concerned about inorganic chemists and my faculty colleagues. Achieving NAS membership begins with a nomination which is then followed by an election—the process is so complicated that, even given the intelligence of our members, most of us do not understand the details of the process—myself included. What is known is that in order to get elected one needs the support of members in more than one subdiscipline of chemistry: analytical, inorganic, organic, and physical. Since there are many more members who are organic or physical chemists, it is readily seen that the playing field is not equal. Understandably, organic chemists generally vote for nominees of its group, which is true for members of each of the other three subdisciplines. This means that the least number of votes are obtained by analytical and inorganic chemists. For many years, only a very small number of analytical and inorganic chemists were elected to membership. However, there has been a dramatic change in recent years, particularly with inorganic chemistry. It has had explosive development in important fundamental research in the chemistry areas of organometallics, bioinorganic, and solid states or new materials. All chemists, regardless of their discipline, are now more aware of who is doing the best research in one or more of these areas. This means that inorganic chemists currently have a much better chance of being elected to the NAS. Likewise, the number of inorganic chemistry members is now large enough to have an impact on the election results.

As stated above, one reason I try to attend all of the NAS annual meetings is to talk with noninorganic chemists about some inorganic chemist(s) who is a strong candidate for NAS membership. One must be very diplomatic in such a conversation, because some members would not want to be told whom to support and for whom to vote. In addition, my informal discussions with other members facilitated my understanding of how they perceived the research of my faculty colleagues. If the response was very positive, I would ask the person if he would be willing to nominate my colleague. I would tell him that I could prepare the nomination, so all that he need do was to edit it and make any changes that he thought were necessary. It would be pointless for me to nominate a faculty colleague, particularly if the area of research differed from mine. Some NAS members would view this as merely an attempt to get one of my colleagues elected as a member. I derive some satisfaction from knowing that I have been successful in a couple of cases by using these two approaches.

Finally, another reason for the unbalanced playing fields is that faculty at certain eastern or western coast schools are more likely to be elected than faculty in the Midwest. It is clear why this happens—these schools have faculties with large numbers of NAS members who, understandably, vote for their own colleagues. When this is pointed out to a member of such a faculty, they usually respond that they have more members because their faculty is known to be better than the faculty in a Midwestern university. Fortunately, there has been a gradual change in the nomination and election of members. There has been a marked increase in the election to membership of inorganic chemists. This is largely due to the present important status of high quality research being done in inorganic chemistry. Likewise, as the number of inorganic chemist members increases, they will have some influence on the other members, thus enhancing the chances of electing more inorganic chemists.

During my many years as a member of the NAS, I have served on different committees, helped prepare desired reports for studies requested by the National Research Council (NRC), and reviewed other such reports, etc. I even chaired a report entitled "An Assessment of Continued R&D into an Electrometallurgical Approach for Treatment of DOE (Department of Energy) Spent Nuclear Fuel."

In 1980, I served as a member of the *Committee on Chemical Sciences* (CCS) for a three-year term. The primary obligation and function of this committee was to consider areas where chemistry makes important contributions, such as agriculture, energy, medicinals, etc. The CCS was to serve as a focal point within the NRC for concerns and issues of the chemical sciences and technology community. The primary function of the NRC is to research specific areas of chemistry requested by the CCS, and to prepare reports on the subject when justified. As a member of the CCS, I participated in its discussions and recommendations.

Since during this period (1980-83) I became President of the ACS, I realized that both CCS and ACS were working toward the same goals: what is good for chemistry and how can it best be used for the needs of mankind. Therefore, I thought that the two should join forces on points of mutual interest, such as preparing reports, chiefly targeted to congressmen, as well as advising the Federal government. I considered that the ACS, with its competent staff working on public affairs related to chemistry, could work with members of the CCS because of their mutual interests concerning the important role of chemistry in our society. I contacted Frank Press, then President of the NAS, and we arranged a luncheon meeting, with each of us inviting four of our colleagues, to consider the feasibility of such a venture. It was generally agreed that this might be more workable and effective than NAS and ACS working independently. I was very saddened that nothing ever came of this. Reasons were provided as to why this would not work: some seemed justified, others not. In my opinion, the ACS thought that they would get stuck doing *all of the work*, while the NAS, with its prestigious reputation and access to Congress, would derive *all of the credit*. Furthermore, the NAS may have thought that its influence might be less effective if it collaborated and shared the preparation of reports with the ACS. Whatever the reason, this is yet another of my "good" ideas that did not fly. For example, I tried but failed to get ACS to produce a public television program on "The Universe of Atoms and Molecules," akin to Carl Sagan's "Cosmos."

In 1983, the CCS was transformed into the Board on Chemical Sciences and Technology (BCST). This Board was to have its membership broadened somewhat, to reflect its additional responsibilities beyond that of CCS. I was appointed to the Board for a three year term. The primary mission of the Board is to inform the general public

Figure 4-3. Professor George Pimentel (1922-89), of the University of California, Berkeley, was an outstanding physical chemist, as well as a great teacher. He was largely responsible for the NAS report *Opportunities in Chemistry*, more often called the "Pimental Report." Credit: ©Andrée Abecassis.

and Congress, in layman terms, on how essential chemistry is to our present welfare and how much it will be needed in future technological developments yet unknown.

The most important activity that I was involved in during my years in the NAS and, particularly, the BCST, focused on the book entitled *Opportunities in Chemistry*, more often referred to as the "Pimentel Report," named after the very able chairman of the report, George Pimentel (**Fig. 4-3**). The book "describes the contemporary research frontiers of chemistry and the opportunities for the chemical sciences to address society's needs." I was privileged to first work on its planning committee which unanimously recommended that the survey of chemistry be undertaken. When the BCST approved, I was on the committee, with 25 eminent scientists, assigned the job of generating the book, which was issued in 1985. This was a real tour-de-force as we often had to meet in Washington, DC, twice a week for a period of about a year. This book was preceded by the book entitled *Chemistry: Opportunities and Needs* (1965), often called the "Westheimer Report" after its chairman, Frank Westheimer. The book was intended for congressmen who should be concerned about science and technology as it often makes significant contributions to the welfare of the people of the US as well as the nation's GNP. It has not been possible to accurately measure the impact of the Westheimer Report, but chemists derived a sense of gratification that the important role of chemistry was available to anyone interested, including other scientists. Physics has also published two books on the contributions of physics, akin to the two in chemistry.

The "Pimentel Report" might never have been written were it not for the keen perception of George Pimentel of the importance of its production. As indicated in its Preface:

The committee's charge was to describe:
- the contemporary frontiers of chemistry,
- the opportunities for the chemical sciences to address society's needs, and

- the resources needed to explore these frontiers to advance human knowledge and to exploit chemistry's opportunities to enhance the well-being of humankind.

For a few years, it was clear that the Pimentel Report was being seriously read. This was particularly true for chemists in the US and other countries and for some layman groups. For example, I was invited to talk about chemistry and the Pimentel Report by the Evanston Lions Club and by the North Shore Senior Center. Both of these groups consist of intelligent citizens who have a general interest in significant topics of which they have little or no knowledge. Believe me, it is not easy to give a talk on science to such groups, but I try my best, hoping to leave them with something that they might remember. One thing I often state is that "were it not for chemistry, all of us here would be naked." I follow this by informing them that most clothing consists of some synthetic material made available by chemical research. This is evident by the fact that it is almost impossible to buy any apparel made of pure cotton. I then stimulate considerable discussion by asking them to name some objects that they come in contact with each day that have not been touched by chemistry. The groups, often composed of Chief Executive Officers (CEOs) of large corporations, bankers, and even politicians find it difficult to do this, but always they manage to give a few examples where chemistry is not involved. Most chemists think it is not possible to give a meaningful lecture to nonchemists. This is not true. What *is* true is that it is almost impossible to have an audience of the general public attend a scientific talk. This means that chemists usually give talks on the importance of chemistry to other groups of chemists which is meaningless, as it is "preaching to the choir."

The response to the Pimentel Report was illustrated by the numerous invitations George received to give talks on it. He called me one day to say that he had invitations to go to the annual meeting of the Canadian Chemical Society and to Japan. He did not think that he could make these trips, and wanted to know if I would be able to go in his place. I said yes, if necessary, but that it was he they wanted to hear and he should make the effort to honor their invitations. I did not know at the time that George was ill and, unfortunately, he died of colon cancer soon afterwards. He did manage to go to Japan, but I pinch-hit for him in Canada. The panel-led discussion was well attended, and there could be no doubt that most of the chemists present had read the report. It is not known if any of our members of Congress or the administration ever read the report that was sent to them. However, we felt that the effort was necessary because if at some congressional hearing the subject of chemistry had to be discussed, the NAS could tell them that they should do their "homework" and read the Pimentel Report.

I am humble but proud to be a member of the NAS. Since its congressional establishment in 1863 as the National Academy of Science to advise the federal government, it has often been called upon to assist the administration and/or Congress on such matters. This has been a step in the appropriate direction, but some of us feel that even more use should be made of the Academy.

AMERICAN CHEMICAL SOCIETY (ACS)

The American Chemical Society (ACS), founded in 1876 with John Draper (1811-82) as its first President, is the largest scientific society in the world, having some 165,000 members. Approximately 620 staff people work at its home location in Washington, DC; another 1,160 are employed at its abstracting and publication site in Columbus, Ohio. One of the most important contributions made by the ACS is its continuing effort to summarize *all* scientific publications—worldwide— having to do with chemistry. *Chemical Abstracts* has published thousands of volumes of abstracts over many years (**Fig. 4-4**). Each year it breaks its own previous Guinness World Record for the most volumes of an index. I vividly recall making good use of *Chemical Abstracts*, painstakingly transferring pertinent information onto index cards, during many hours I spent in the library when collecting information for the Basolo-Pearson book. Now, in the age of computers, one only has to input a significant scientific word, and the computer performs a search of the literature. In a short time, with little effort, one receives a complete list of the literature containing the key word that can be printed out, or the customized list can be further reduced by adding another key word.

I believe that the two most important global contributions that the ACS has made for chemists are its long list of journal publications and the publication of *Chemical Abstracts*. The *Journal of the American Chemical Society* (*J. Am. Chem. Soc.* or *JACS*) is the very best, or one of the very best, chemistry journal(s) worldwide that publishes original scientific research on any aspect of chemistry. Chemists engaged in research know they must read *JACS* in order to keep up with the cutting edge of important advancements. Therefore, chemists almost always submit their very best work to *JACS*, because they know that the paper will most likely be read by chemists in general, rather than only chemists who are specialized in a particular area of research.

In addition, the ACS publishes specialized journals such as *Inorganic Chemistry*, *Organometallics*, and many others. As science and technology move forward and a need for a specialized journal is perceived, the ACS will most certainly produce it. At present, the ACS publishes 35 journals reporting on original research in chemistry. *Biomacromolecules*, *Crystal Growth and Design*, and *Nanoletters* are

Figure 4-4. Each year, the American Chemical Society publishes the *Chemical Abstract Index* of all papers published with reference to chemistry. It always holds the Guiness World Record for the longest index, as shown here in 35 volumes.

the most recent journals that have been added in order to cover the current hot areas of chemistry. Finally, *Chemistry and Engineering News* (*C&E News*) provides the weekly chemistry news. I think it fair to say that *all* chemists who receive this news magazine read it, or at least browse through it.

The ACS also serves several other important functions for chemistry and chemists. These need not be discussed, but I would be remiss if I did not at least mention the two annual ACS meetings, each attended by about 15,000 chemists; its departments of education, public relations, and foreign affairs; and many committees.

My first involvement with the ACS began in 1941 when I was a graduate student. As a student member, I received lower rates for membership and for *JACS*. During the remainder of my student days and my years at Rohm and Haas, the only contact I had with the ACS was to attend a few of its annual meetings.

My first active participation with the ACS was in the mid-50s when I was invited to make a lecture tour of local ACS sections in the Pacific Northwest. I got excited about the tour because it came at a time when our research on ligand substitution reactions was rapidly producing some important results. I still recall a few things that happened during this lecture tour, the first of many which followed over the years that I will not mention. This first tour of the Northwest Local Sections of the ACS scheduled me at Spokane, Billings, Butte, Missoula, Seattle, and Pullman—all in one week, so I was on an overnight stand at each. At each stop, it appeared that all the section members were in attendance, and it also seemed as though they enjoyed getting together once a month, regardless of the speaker or topic.

I do not recall the order of my stops, but I think the first was at Spokane, Washington. Because of my enthusiasm about our latest "hot" (the young chemists now say "cool") results, I was eager to tell anyone who would listen about our findings. However, at the end of my first talk, there were only two questions:"What is a ligand?" and "What is a metal complex?" Needless to say, I was very disappointed because it was clear they had not understood what I wanted to tell them about our research. That night I wondered why my talk was such a failure, and what I should do to try to correct it for my remaining lectures. I determined that it had to be mostly my fault for not having properly judged the familiarity of the audience with this area of chemistry. I knew of the saying that "one should not underestimate their audience" when giving a talk. Now I learned it is equally important "not to overestimate the listeners." The remainder of my talks improved because I spent a great deal more time on the introduction than on our research.

The other thing that I recall about my first talk at Spokane College is that one of the faculty escorted me to his lab to show me how he had prepared perfumes, all having different odors. He insisted on giving me samples of each to take to my wife. When I arrived home, I gave them to her. She smelled the various perfumes but never used any, and eventually they were thrown out. I think that the college in Spokane is Catholic, so it is probable that the perfume chemist was a priest. I wrote him a kind letter informing him that the perfumes were a big success with my wife. I trust his divine powers did not convey to him that I had told a little white lie.

This first tour also taught me an important lesson which was to make certain that the last slide is not left in the projector. In those days the slides were large (2 by 3 inches) and had to be handled by a projectionist. At each stop, my last slide would be forgotten in the projector and, as a result, my talks became shorter and shorter!

Some six months after my trip to the northwestern ACS local sections, I was asked to do the same for eastern sections. One of the stops on that tour was Washington, DC, where they always had two speakers who would give their lectures at the same time. One of the talks was usually on a topic in organic chemistry and this speaker was given the bigger lecture room to accommodate the larger number of organic chemists. However, on this occasion, my talk attracted the larger attendance, and there was standing room only in our smaller lecture room. This occurred because many of the organic chemists chose to hear my talk rather than the organic talk. The question was: why should this happen? The organic chemists told us that they had read a recent article by Sir Christopher Ingold and Sir Ronald Nyholm attacking our proposed S_N1CB mechanism for the base hydrolysis of Co(III) ammine complexes. Organic chemists knew Ingold as one of the world's most outstanding physical organic chemists. Since Ingold published that the mechanism, proposed by two young assistant professors was wrong, the organic chemists came to hear why the inorganic chemists were wrong. A few of them came to me after the talk and said that the experimental facts we had collected were consistent with our mechanism, and it appeared that the S_N2 mechanism proposed by Ingold might be wrong. Subsequently, many experiments were carried out worldwide which showed that Ingold's mechanism was indeed in error and that our mechanism seemed to be correct (pg. 88).

My next interaction with the ACS was in 1968 when I served on the Petroleum Research Fund Committee (pg. 144).

I also served on the committee that visited departments of chemistry which had not yet been accredited by the ACS. The minimum standard that a department had to meet for accreditation was to offer courses in the four different disciplines of chemistry and to have adequate labs with necessary ventilation, chemicals, and glassware. More often than not, the departments that we visited fell short on some of these items, usually due to a lack of resources. At many of the colleges, members of the chemistry faculty had repeatedly and unsuccessfully requested funds from the administration in order to take care of some of the needs of the department. Since improvements had not been made, the department could not receive ACS accreditation. Before leaving, members of our committee would have a long talk with the President of the small college. The usual response that we heard was that the school depended almost entirely for its funds from tuition paid by the students. After the faculty salaries were paid, this left very little to be of much help to the departments, and it was less costly to provide funds to departments other than the more expensive science departments. We would then point out that good science courses in colleges are needed if a science major wants to go onto graduate school and obtain their Ph.D. Furthermore, if small amounts of funds were set aside for the sciences each year, as is done for libraries, it would gradually improve the department to the point where it would be able to obtain ACS accreditation. We did not say anything different to the college president than what he was being told by the chemistry faculty, but the major difference was that our statements had greater credibility, assuring the President that the requests being made by the department were justified. I especially liked being assigned to this committee because of my interest in teaching. It was rewarding to have one of the faculty tell me how much they appreciated our visit because the administration, now convinced the department needed help, was concentrating on providing that help.

Another early contact I had with the ACS was in the late 40s and early 50s, when we had done enough research to have me present 15 minute talks at the semi-annual meetings of the ACS. At that time, posters to display and discuss one's work were not yet in use. Few symposia were being held at those meetings, making for good attendance at the 15 minute presentations and 5 minute discussions. I vividly recall the talk that I almost failed to give at the Chicago meeting of the ACS in 1960. I know this was the date, because my daughter Liz has just had her 40th birthday. My wife had spent much of the night and wee hours of the morning in labor, with me awake in the waiting room. I was concerned about the birth of our child, but I also knew I was scheduled to give the opening talk at a session chaired by John Bailar, which started at 8:30 AM. It seemed as though I would not be able to arrive in time, so I telephoned Bailar in his hotel room to tell him of my predicament. Elizabeth arrived at about 6:00 AM. I waited until about 7:00 AM when the nurse assured me that everything was OK to leave. I quickly took the commuter train to Chicago, and ran to the hotel room where the presentations of our group were being held. I was puffing from running, with slides ready to give my talk. Bailar was in the process of explaining to the audience why he would have to cancel my talk. The attendees saw me arrive before Bailar did, and they started applauding, making John wonder why his announcement prompted such a reaction. In jest, he later told me that he thought they were applauding because they were so happy that I would not be there to give my talk.

In 1970, I was elected Chairman of the ACS Inorganic Division and will first provide a little history and some comments about it. The Division of Physical and Inorganic Chemistry was formed in 1908. The number of inorganic chemists at the time was too small to have its own division. The number of physical chemists grew more rapidly than the inorganic chemists and, within a few years, the division consisted largely of physical chemists. Naturally, the division programs reflected the interest of this predominant group. As a result, the inorganic chemists began to feel like second-class citizens, and, as this situation worsened, the possibility of forming an Inorganic Division began to emerge. The senior inorganic chemists (J. C. Bailar, Jr. (1904-91), H. S. Booth (1891-1950), W. C. Fernelius (1905-86), and T. Moeller (1913-97)) sent letters to all the inorganic chemists they knew. The letter outlined the advantages and disadvantages of having a separate Division of Inorganic Chemistry, and asked for a response by voting yes or no. Of the 50 letters sent, the vote was almost a unanimous yes in favor of the division. The fifty who received letters sent copies of the letter to inorganic chemists that they knew. In a few weeks, there were about 150 pledges. This was then submitted to the Division of Physical and Inorganic Chemistry, requesting the formation of a Division of Inorganic Chemistry. This was granted with the stipulation that the physical division keep all of the funds of the division, and the inorganic chemists agreed. In 1957, the Division of Inorganic Chemistry was approved by the ACS. The members numbered 476 at the beginning, 2,860 in 1982, and 5,908 in 2000. This shows the growth of the membership, which is a reflection of the increased research activity in inorganic chemistry. It now has the largest number of research displays at the semi-annual meetings of the ACS. It is believed that growth will continue as more and more graduate students choose to do their dissertation research in inorganic chemistry. One other reason for its growth is that industries continually demand many new products that can only be produced by inorganic chemistry.

My involvement with the Division of Inorganic Chemistry took many forms, which ended with my becoming Chairman in 1970 and a member of the executive committee in 1971. Neither of these appointments required a great deal of time. When I was chairman, I had help from the chairman-elect, the secretary, and the members of the executive committee. My duties were merely to keep in touch with the persons involved to make certain our division had papers to be presented at the semi-annual ACS meetings, and at the special biannual division meeting. The only thing other than this of any note was our attempt to get the Dow Corning Corporation, which sponsored the Frederic Stanley Kipping Award in Silicon Chemistry, to change the award to include all metals. There were two reasons for making this request. One was that we did not think there was enough high quality important research left to do on silicon. The other reason was that there should be an award for organometallic chemistry which was just blossoming and surely destined to grow. Somewhat as expected, Dow Corning, the manufacturer of silicon products, said no, that there was much more good research to be done on silicon. Time has proven them to be correct. We, too, were correct, and there are now two journals of organometallic chemistry: *Journal of Organometallic Chemistry* and *Organometallics*. Now there is also the ACS Award in Organometallic Chemistry, sponsored by none other than the Dow Chemical Company Foundation. An example of the importance of this research is the use of metallocenes as homogeneous catalysts to manufacture tons of polyethylene daily. That other important discoveries will be made in this area of chemistry is assured by the fact that research with such compounds is being done by outstanding young chemists who will continue to propagate such developments.

My final significant interaction with the ACS was when I became its president in 1983. This was not planned by me, nor did I do any electioneering to be elected. Two members of the ACS had been nominated. Petition nominees could be added, provided they had a sufficient number of supporting signatures. One evening, I had a telephone call from one of my very best friends who was extremely credible and honest. He explained to me that he and many others were not satisfied with either of the two nominees becoming president. They wanted to get the necessary signatures and propose me as a petition candidate. My friend was prepared— three more of my friends were present and each got on the phone to ask that I please be willing to run for president-elect of the ACS. I told them that should I run I would not raise a finger to electioneer in an

Figure 4-5. Mary and me in 1983, during my year as President of the ACS.

attempt to win. They said that was OK and they would do all the electioneering. I finally said I needed to think about it, and they should call me again in three days when I would give them my decision. I discussed this with my wife Mary and she said, "I know your friends who ask you to do this. I think if it will be good for chemistry which you like so very much, you should do it." I told her this would require that I travel a great deal so I would have even less time to spend with her and the children. She made the point that the children were old enough to take care of themselves. Mary also said, "It may even be of interest to me, for I can arrange to have the children looked after and do some traveling with you."

Therefore, I was prepared to say yes when my friends telephoned me. They were very pleased and said they had already obtained enough signatures for my nomination. I cautioned them that their efforts may go for naught, because the other two candidates were part of the loyal ACS establishment and would get all of the establishment votes. Since I had not been very active in the society, I could well not win. They remained optimistic and assured me I would be elected. They were correct, and I was elected President-elect to become President the following year (**Fig. 4-5**).

I was sure that I would be defeated because the nominees were required to write an article for publication in the weekly magazine *C&E News*. One of the questions we had to address was: "What do you wish to accomplish as President?" What I wished to accomplish was 180 degrees opposite to that of the establishment. I wanted the society to experiment with holding only an annual meeting. The need for two meetings to quickly report some important chemical research was no longer valid, due to the large number of national and international chemical conferences and symposia. I also suggested that a review of all of the committees was in order. Although there is often a reason for appointing a committee to handle a problem that needs attention, when the job has been done, the committee should be terminated with kind thanks for having helped. It is easy to form a committee but not so easy to remove it when its mission has been completed. Finally, I was of the opinion that society by-laws should put an upper limit—10 or 15 years—on the total number of years that one could serve on committees at the national level. At present, some of our members work hard and do a good job, so they do not understand why their good work should be terminated at some upper number of years. Some have indeed done good work for 20 to 30 years, but one thing they have not done is step aside to give other ACS members a chance to serve. Ten or fifteen years on national level committees provide ample time to promote views and prompt changes in the society. Then it is time for new people with fresh ideas to serve on these committees. In my opinion, this is common sense, and other organizations must think the same as they most generally have a limit of 5 years plus renewal of another 5 years as the limit of service on committees. My biggest disappointment during my Presidential year was that I was not able to make any progress with the idea of one national meeting a year or the problems with committees. That shows the little I knew about the entrenched establishment at the ACS.

Fortunately, there were many things I did that I felt good about during my Presidential year. What follows is a chronological mention of what I recall most during the year. First, Mary and I enjoyed all of the events, of which there were many, that she was to attend with me. I will always remember her excitement during the ACS meeting in Seattle. A large limousine was awaiting us at the airport. It had a big back seat area with a bar and a small TV. At the hotel we were given the President's suite with a beautiful view of Puget Sound. The living room, bedroom, and washroom were all large. However, the most excitement was caused by a large sunken bath tub in the center of the larger washroom. This was so unbelievable to Mary that during the week when she met someone she knew, she would invite them to our suite to show them the bath tub. At some point I said to Mary, "Who would ever believe that some poor unfortunate kid from Coello would ever make it here?" We both had a good laugh, but we admitted that we enjoyed it.

Mary also came with me to a gala reception in Washington, D.C. This reception is held each January to invite congressmen to meet the ACS President and other officials. I was somewhat surprised to see that several of the congressmen did attend. This became understandable when I saw the huge amount and variety of goodies and some of the best wines on a large long table in the center of the room, where guests could serve themselves. Mary was spellbound by the mountain of shrimp in the center of the table. This was her favorite hors d'oeuvre and at most receptions she only would have a few shrimp before they were all gone. This time she had more than her fill of what she liked most from the bounty on the huge table. I was pleased to have an opportunity to talk with Congressman George Brown of California. He, as congressman, always supported legislation in favor of science and technology. In most cases, it was he who

Figure 4-6. Arnold Thackray is responsible for establishing the Chemical Heritage Foundation. Mainly through the efforts of Thackray, who understood the significance of chemical achievement, and who was very adept at raising large gifts from generous philanthropists, the center is now recognized globally as an excellent resource for the history of chemistry. The Center is located at 315 Chestnut St., Philadelphia, PA 19106-2702, USA. Credit: The Chemical Heritage Foundation Image Archives, Othmer Library of Chemical History, Philadelphia, PA.

introduced bills that asked for funding of scientific research. It was a sad moment for chemistry when he resigned.

My first duty as President-elect of the ACS was to spend the year getting familiar with the workings of the society. My first real job was to appoint and/or remove members of committees. I was assisted in this by a member of the staff who had the responsibility of overseeing the operations of committees. I did not know most of the persons involved, so I went along with the recommendations of Halley Merrell. At the end of our consideration of the individuals involved, we had not made many changes, except to move some people from one committee to another because there is a rule (or by-law) which limits service on a particular committee. Unfortunately, this does not prevent a member from being transferred to another committee, and this shifting from committee to committee makes it possible for a member to serve the society at the national level for 20-30 years, if desired. There were a few persons on the list whom I knew, and did not want to reappoint. Halley informed me that I was rocking the boat, which would not be taken lightly by any of the many committee members. I stood my ground, but he was right in what he had told me.

As I progressed with my on-the-job learning as President-elect, I was asked to attend a meeting of the Board of Directors. It was understood that I had no voice nor vote at the meeting. After listening to the Board deliberations for two days, an item on the agenda that needed to be discussed was that of supporting an endeavor to establish a center on the History of Chemistry at the University of Pennsylvania. The person, Arnold Thackray (**Fig. 4-6**), who was attempting to get funds from the ACS was asked to come state his case of why such a center was important to chemistry, and why the

ACS should help fund it. Thackray made a good presentation of what he planned to do and why it was of importance to chemistry. After he had departed, the Board members discussed the issue at length. At some point I asked permission to make a comment, and this was granted. I stated that, of all the items on the agenda and the considerations given them, I felt that the center for the history of chemistry had a chance of making the most impact on chemistry over the long term. My remark had good impact, for members of the Board had been quite ambivalent over the past several months as to whether to help such a center get its start. Now, at last, they made a positive decision. Funds were contributed to the effort, and more were later provided by the American Institute of Chemical Engineers. Through mainly the efforts of Thackray, who understood the depth and significance of chemical achievement—and who was also very adept at raising large gifts from generous philanthropists, the center was established and is now recognized globally as an excellent resource for the history of chemistry.

Figure 4-7. Me in 1983 as President of the ACS. My official capacity, at the national level, was one year each: President-Elect, President, and Member of the Board of Directors.

I became President of the ACS on 01 January 1983 (**Fig. 4-7**). Almost immediately, I began my crusade to establish only one annual meeting per year, terminating all committees no longer needed, and changing the by-laws to put an upper limit (10 or 15 years) of time that a member could serve on national committees. Members of the ACS know how I failed miserably with these three goals.

One of the duties of the ACS President is to represent its positions. Since I have just been told that I have been selected to receive the Priestley Medal[*] in 2001 (see **Fig.**

---

[*]It can be said that chemistry, as we know it today, started with Joseph Priestley 1733-1804) and his adversary Antoine Laurent Lavoisier (1743-1794). Priestley is noted for his discovery of oxygen, but he also discovered hydrogen chloride, sulfur dioxide, and nitrous oxide. He retained his belief in his phlogiston theory until his death. This theory was finally proven wrong by Lavoiser in the late eighteenth century. Priestley also coined the word "rubber" when he noticed that couchouc would erase pencil (graphite) marks. As Ralph E. Oeper notes in his book, "The Human Side of Chemists," Priestley was an outspoken dissenter. This led to his antagonizing the clergy and adherents of the Anglican Church. As a result, his house in Birmingham was burned, so Priestley and his family came to the US. He bought a large tract of land in Northumberland , Pennsylvania, and lived there the rest of his life. The house is open to the public and one can see some of the pieces of apparatus that were used by Priestley. In honor

7-6), I will start with my representation of the ACS at Priestley's birthday celebration in Northumberland, Pennsylvania.

The birthday celebration was held outdoors on a beautiful sunny day at the Priestley home. A long, high board where the speakers would sit was positioned in the yard. Each speaker gave a five minute talk in the order of their seating, left to right. People attending the celebration simply stood in the yard. In order of seating, the speakers were the town mayor, the town postmaster, a representative of the US postmaster, others, and I. Before the talk, there were about 100 persons present, milling around, having coffee or tea, and talking with one another. Almost all of the attendees were members of the Unitarian Church, who knew Priestley as one of the founders of their religion, as well as an outstanding chemist. Since no chemists were present, it occurred to me that I had a captive audience of nonchemists who should hear about the good side of chemistry. There were three speakers before me, so I had about 15 minutes to think about what I should say, rather than what I had planned to say.

What I said was that, on this his birthday, Priestley would be very pleased and proud of what chemistry has achieved. He would see that chemistry provides us with essential things in our everyday life, such as toothpaste, clothing, detergents, etc. as well as several much larger and more important endeavors. These would include agriculture, with its use of fertilizers and insecticides to increase the production of food, improved medicinals and the discovery of new and effective drugs, oil refinery for gasoline production, petrochemicals for use in the manufacture of plastics and many other products. Yes, Priestly would be proud to see what chemistry has achieved in the past 200 years to improve the welfare of mankind.

After all the talks had finished, people stayed a little longer to go through the house and to visit with one another. During this time, four men came up to me and introduced themselves. One said, "We knew Priestley was a chemist, but we were surprised to have a chemist come to represent the ACS. We keep seeing in the news media how chemicals such as dioxan and others are toxic or carcinogenic. We have gotten the opinion that all of chemistry is bad and we even expected that chemists wanted not to be known as chemists. It was good to have you come to tell us how chemicals are helpful in our everyday life."

The handling of dioxan and other toxic or carcinogenic chemicals was a particularly newsworthy item requiring the response of the ACS during the year 1983. As president, I represented the ACS and was often asked by newspaper reporters, radio talk show hosts, and television anchor persons to discuss the position of chemists on these environmental problems. I agreed to do this but always with one caveat—that I be given equal time to discuss the beneficial aspects of chemicals. Some of the interviewers hesitated, but after determining that I really meant it, they would say

---

of his birthday celebration, John Bailar and the ACS were able to convince John's son Benjamin Bailar (then US. Postmaster General) to issue a Priestley postage stamp. A stamp issued in honor of a famous scientist is most unusual in the United States, although it is often done in foreign countries.

OK. Some would even request questions that might be used about the good contributions that chemicals bring to our society. I provided them with the questions, and the last half of our conversation on the good aspects of chemicals often proceeded rather well.

I would start this part of the interview with the point that 99.9% of all manufactured chemicals are good for us, whereas only the remaining 0.1% may cause us problems. I further made the point that the reason many people think *all* chemicals are bad is because reported news focuses on the bad aspects of *all* problems—for example, the daily news reports on the bad things we do to one another, such as killings, rape, etc. Rarely do reporters cover a positive story and extend it for weeks. At this point, the host would start to ask me the questions which I had provided him. The first question was "How does chemistry improve our everyday life?" I would then state that much of what one takes for granted and comes in contact with during the day is a product of chemistry. Let's say the day starts with brushing your teeth and ends in taking a sleeping pill. These and most of what you have made use of during the day are products resulting directly or indirectly from the research of a chemist. For example, most toothpastes contain fluoride because it hardens tooth enamel and helps to prevent the formation of cavities.* I also would state that doctors who practice internal medicine would be in serious trouble if research and medicinal chemists had not synthesized and discovered prescription pharmaceuticals.

The next question was: "Why, if chemistry provides us with all these useful commodities, does it cause so much pollution?" In response, I would state that this harmful effect on the environment is frequently the result of the formation of hazardous side products, and that the chemical industry has worked very hard to negate the formation of such by-products. Furthermore, the government (OSHA) bans the use of carcinogenic and toxic substances. It is clear that harmful chemicals such as dioxan must be eradicated, and it is likewise clear that chemists will be the ones to do this and, in general, to clean up our environment. For example, chemists at du Pont have developed a chemical to replace Freon® (CCl_2F_2) which is, in part, responsible for the depletion of ozone (O_3) in the atmosphere. Other chemists are doing research on making a smaller, more efficient battery for use in automobiles that can lessen the formation of carbon dioxide (CO_2), which is said to be causing global warming.

In a half hour interview, we often had time to likewise talk about the contributions of chemistry to agriculture and energy. One final point that I always tried to make clear was that accidents with harmful chemicals are often not the fault of irresponsible chemists. The manufactured product is put in large safety containers and transported by trains or large trucks. Many shipments are made daily without incident, but, at times, a train or a truck has an accident. This may break an otherwise safe container, releasing

---

*This discovery was made by Professor William Nebergall (1914-78) at Indiana University. The chemical initially found to give the best results was stannous fluoride (SnF_2). Sodium flouride (NaF) is most commonly used today.

the harmful chemical. When this happens, the news media always blames the chemical whereas the blame should be put squarely on the train or the truck.

Unfortunately, these interviews may have had little effect on the general public. Even a half hour TV interview does not seem to hold the interest of the nonscientific viewer. For example, my first interview was taped and I was given a tape to show my family. My adult children and grandchildren were all excited about wanting to see me on TV. What happened tells me a lot about the effectiveness of such interviews on chemicals and chemistry. After the first 5 minutes, one grandchild left to go to the washroom, another followed, and in less than 10 minutes none of the five grandchildren were watching grandpa on TV. One of my children left after 20 minutes to care for the playing grandchildren. Only one of my children watched the entire half-hour interview. This sadly informed me that if my family members cannot stay focused on their dad or granddad on TV for half an hour, imagine how quickly others would switch the channel on their TVs!

The President of the ACS also has the duty to represent its viewpoints before some congressional committees. I had to do this twice: once before a committee wanting to know the ACS view on the teaching of science and math in our public schools (grades K to 12) and once before a committee on the funding of basic research. The ACS staff prepares a brief to be entered into the record, and the committee members ask a few questions that generally are easy to answer. The disturbing feature about such presentations is that there are so few committee members present. They come and they go, but the number at any one time is consistently small. What *is* perhaps useful is the report of the ACS left as a record of its position on the subject matter. Unfortunately, even the report may never be read by a congressman.

The meeting often ends at noon, followed by lunch. One or two congressmen may be present, but more often staff members are all that remain. At one of these lunches, the conversation got around to the environment. One of the staff members, quite loudly, kept saying that the environmental problems were due to chemicals. He went so far as to say that "Chemicals are bad. We must get rid of all chemicals." During our luncheon conversation, he repeated this once or twice. At some point I asked him where he and I would be if we got rid of all chemicals. He had no meaningful answer, so I told him we would not be here because our bodies are made of chemicals. True, the biological chemicals and their molecular reactions are complicated, but chemists, along with medical doctors doing fundamental research on these systems, continue to learn more about the body's secrets. I went further to say that all things on planet earth are made of chemicals, including earth itself. The only things that are not chemicals are sound, light, heat, and magnetism. I left thinking that the staff members had failed to get the message.

As President of the ACS, I also had two other responsibilities: representation of the ACS with (1) the White House Science Advisor, George A. Keyworth, and (2) the Honorable Donald Hodel, Secretary of Energy, US Department of Energy (DOE). I arranged an appointment with Keyworth who had announced (1983) the final assembly of a nearly full complement of top aides at the White House Office of Science and

Technology Policy. Bob Parry and I met with him in his office to make the point that the ACS believed a chemist should be among the top aides appointed to give advice on science and technology policies. The group that Keyworth had appointed consisted of 3 physicists, a management Ph.D., a mechanical engineer, and a systems manager. Bob and I wanted to stress the issue that a chemist should also be on this list, because of chemistry's direct impact on agriculture, energy, health, and the US GNP. Keyworth was a good listener, and finally he agreed that this was an oversight but he did not promise to take care of it. Sorry, but I do not recall if a chemist was ever appointed and, a few years later, Keyworth resigned his position.

My correspondence with Hodel on behalf of the ACS concerned chemistry's direct role with energy on planet earth. Petroleum refineries give us high octane gasoline. Sulfur removal from coal permits its use with little effect on the environment. The concern of the ACS was that the scientific community, in particular, the chemistry research community, needed additional avenues for formal input into the Office of Basic Energy Sciences of the Department of Energy. The Society, therefore, recommended that the DOE establish an advisory committee on the chemical sciences to its Office of Basic Energy Sciences. Some of the duties of the committee would be proposal review and evaluation, oversight of program management, overall program balance, etc.

It is satisfying to know that DOE now supports basic research in chemistry, particularly that related to its mission. Perhaps the ACS contact with Honorable Hodel was somewhat useful.

The ACS has an annual breakfast to which it invites congressmen so that they may meet with its President and members of its board. Again, the congressmen do not usually attend, but some of the staff do come. One member of the House of Representatives, Albert Gore, was there and I was pleased to have him sit at our table. He struck me as being an ambitious young man, anxious to move ahead quickly, and make a name for himself. He had chosen the issue of the environment to help him there, so he said he was glad to have the opportunity to discuss his views with us chemists. I think most of us at our table came away thinking that he was not yet ready to tackle the problems of the environment. I hope he has had sufficient time with on-the-job learning to tackle the task of possibly being our next president (written before the election, which we now know he lost).

My year as President of the ACS was a rewarding experience for me. I learned that one can not succeed in making major changes in an organization deeply entrenched in its policies. However, most of the effort I made as President was a good learning experience for me, and some of what was done made significant contributions to chemistry. Therefore I take this opportunity to thank the members of the ACS who voted for me to be their President in 1983.

GORDON RESEARCH CONFERENCES (GRC)

The Gordon Research Conferences (GRC) are named after its founder Neil Elbridge Gordon (1886-1949). He was a chemist with a Ph.D. degree from Johns Hopkins University. Some of his colleagues were more esteemed as chemists than was

Gordon. However, he did something for science, and chemistry in particular, that remains unique today. He conceived a lecture series with open and uninhibited discussion of key scientific issues. The first two lectures in 1931 and 1932 were held in Remsen Hall at Johns Hopkins. In 1934, Gordon was able to move the conferences to the private Chesapeake Bay vacation haven, Gibson Island. The conferences' next move was to New Hampshire in 1947, using New London as its first site. This was the same site used for several years by the inorganic chemists. From 1947 to 1968, George Parks served as director of the GRC.

I would be remiss if I did not acknowledge that the GRC was able to attract about 55 corporate sponsors to provide initial funding. The founding sponsors no longer need to continue funding. Although the inorganic conference started in 1951, I did not get to know Parks. He was described by some as "imperial" and "autocratic."

Figure 4-8. At bottom right is Neil E. Gordon (1886-1949), the founder of the GRC. Going clockwise are the directors in chronological order: George Parks (1904-75), Alex Cruickshank, and Carl Storm.

He was fortunate in having a genial and able assistant director, Alexander Cruickshank (**Fig. 4-8**). All of my work regarding the GRC was done with Alex and we became very well acquainted and the best of friends. One thing that we had in common and could immediately talk about was Professor J. Harold Smith, who had been on the faculty at the UI during my years there as a graduate student. Smith then moved to the University of Massachusetts, where Alex became one of his graduate students.

In 1993, Alex finally retired, after having devoted himself totally to the Directorship of the GRC. This became part of his life, and he did such a superb job that he will be a hard act to follow. However, Dr. Carlyle B. Storm (**Fig. 4-8**), who succeeded Alex, has gotten off to a running start. Carl retired from Los Alamos to accept this position. Almost immediately the GRC office went from the low tech used by Alex to the high tech of Carl, putting GRC on the World Wide Web (WWW). Also, since about 25% of the participants are from outside the US, it was decided to initiate a GRC abroad. This has been done, and conferences have already been held in the following countries: Italy, Germany, Switzerland, Czech Republic, France, Japan, Singapore, Beijing, Hong Kong, and the UK. Furthermore, conferences are no longer limited to chemistry, but may be on other scientific topics.

The GRCs are unique because a group of scientists with similar interests "live" together for one week, isolated from outside interference. The day starts with morning lectures and discussions. The afternoons are free, and, after dinner, there are evening lectures and discussions. The number of participants is not to exceed 150. Almost all attendees stay the entire week, which gives them ample time to meet one another and, at times, even discuss some chemistry.

During the afternoon free time, some scientists get together and talk about their research, even, at times, preparing a joint paper. Most take advantage of the free afternoon to unwind and partake of some physical activity, such as softball, volleyball, sailing, and, my particular choices, golf and tennis. I was not very good at either of these two activities, but my preference has always been golf. I first "played" a game of golf at age 25 when I was working at Rohm and Haas. This occurred because I agreed to substitute for one of the foursome in our lab who could not play that weekend. I had always kidded my colleagues that golf was a sissy game because anyone should be able to hit a ball which is not moving. After all, I had played a lot of baseball where the ball is moving when one tries to hit it. We were to play golf on Saturday, so my friend Bob Brosious suggested that we go to a driving range and hit a bucket of balls on Friday. I said, "OK, if you think it is necessary." We went and I was at a total loss. I held the golf club as I would a baseball bat, and I took a hard swing at the golf ball. Nothing happened, as the ball remained on the tee. I repeated this a few times with the same results, until I at last made contact with the ball. Even this contact was a disaster, because the ball did not go very far and was certainly not straight down the fairway. Bob then said, "Let me show you how to hold the club and swing it at the ball." He showed me, which helped only a bit. The next day I was embarrassed and totally destroyed by my performance. I had never been confronted with a game that I played so poorly. I decided to take lessons with a pro and became able to play a good game with the average golfer. Now, as a retired Emeritus Professor, if I were not now handicapped, I would be playing two or three times a week.

Before saying more about how the inorganic GRC got started, I will indicate the status of inorganic chemistry in the US at that time. As mentioned earlier, there were very few of us doing research in inorganic chemistry in the late 40s and early 50s. Most of the research was being done in Germany and reported in German journals. Needless to say, most of us Americans did not know German, and even a German-English dictionary did not provide enough help. However, when a German article was really necessary for our research, we would manage to translate what was needed from the article. We found that almost all of their papers described the syntheses and reactions of new inorganic compounds. They *never* reported any detailed studies on why and/or how the reactions took place. For this reason, several of us chemists referred to the German inorganic chemists as *"cookers."*

The nascent period of inorganic chemistry in the US can be said to have started after World War II, when some chemists who had worked on the Manhattan Project returned to their previous positions of employment. The research done by chemists on the Manhattan Project was largely on the fission products resulting from nuclear reactions. Regardless of the type of chemist (analytical, organic, or physical), their research had to deal with metals, which means inorganic chemistry. When returning to their labs, some of these chemists chose to continue their research in inorganic chemistry. In addition, instrumentation such as x-ray diffractometers, spectrometers, and, very soon after this, nuclear magnetic resonance (NMR), etc. were introduced. Such instruments and tracer elements are necessary to do modern research in inorganic chemistry. All of these factors came along at about the same time and, in my opinion, created the renaissance period of inorganic chemistry in the US.

It was then that a few of us decided to get permission from the GRC to start a conference on inorganic chemistry. This request was granted with the understanding that we were to maintain a minimum number of 100 attendees in order that the schools involved could make a profit. A few of us wrote many letters to chemists we believed would want to attend and support the conference.

The first inorganic chemistry conference was held the week of August 20, 1951, at New Hampton, New Hampshire. The attendees numbered 60, with 28 from industries and other nonacademic positions. Considering that we had yet to reach Malcolm Dole's "golden age of America," the academic people had to pay for their own travel and registration expenses. This is probably why there were so few academic scientists in attendance. This has changed dramatically over the years, because different governmental agencies now support the research of many university faculty. Within the budget, a certain amount of money for travel to scientific meetings is always included, making it possible for them to attend meetings of their choice. This means that, generally, persons in academic positions have come to outnumber the industrial chemists at most meetings.

The first inorganic GRC (1951) was chaired by Professor Conrad Fernelius of Pennsylvania State University. His research dealt with measurements of metal complex stabilities in aqueous solutions. The GRC topics were as follows: (1) discussions on crystal growth chaired by H. C. Kremers (Harshaw Chemical Co.) and A. N. Holden (Bell Laboratories); (2) coordination chemistry, chaired by J. C. Bailar, Jr. I gave the first talk entitled "Stereochemical Changes in Reactions of Coordination Compounds" to begin what is now the 50th year of this annual conference meeting. At the time, it did not occur to me that I was giving the very first talk of what was to become the inorganic GRC as it is today. My talk was followed with talks by Bailar, Art Martell of Clark University (now at Texas A&M), and Bodie Douglas (University of Pittsburgh). The third topic was arranged by Anton Burg (University of Southern California), who had earlier distributed outlines to promote open discussions. The discussion was largely carried on by Burg and R. Parry of the University of Michigan (now at the University of Utah). Topic (4) was chaired by Professor W. C. Shumb (Massachusetts Institute of Technology). The speakers were George Cady (University of Washington), Martin Kilpatrick (Illinois Institute of Technology), Helmut Haendler (University of New Hampshire), and John Gall (Penn. State Manufacturing Company).

Chemists, particularly inorganic, looking at the starting GRC program will realize that there were no topics on solid state inorganic chemistry, new materials, organometallic chemistry, bioinorganic, mechanisms of reactions, nanotube chemistry, etc. This should be enough to show the strides that it has made in the past half century. In fact, there are now separate GRCs for each of the areas of inorganic chemistry listed above. This causes some inorganic chemists to say we should no longer have an inorganic GRC, because it is covered in all of the special areas listed above. Some of us old timers (with help from some of our youngsters) keep making the point that there still is reason for continuing the inorganic chemistry conference. It brings together chemists of different interests, and it tries to cover topics not included in the more specialized conferences. A need for this more general conference is shown by its having had more applicants than its required number of conferees for half a century. For example, Professor Kim Dunbar (**Fig. 4-9**), the chairperson for 1999, had 180 applicants and could only accept the limit of 150. The same was true with Dr. Nadine DeVries (**Fig. 4-10**). Furthermore, the conference has been held each year, except once when its chairman became ill and was not able to arrange a program.

I have attended all but about a half dozen, missing two when I was in the hospital due to back surgery and pneumonia, respectively. On both occasions the attendees sent me two large "Get Well" cards with all of their signatures.

Figure 4-9. Professor Kim Dunbar, at Texas A&M University, was the Chairperson of the 1999 GRC. Her research is at the interface of materials and biological chemistry.

Figure 4-10. Dr. Nadine de Vries was the Chairperson of the 2000 GRC. She started her career as a research chemist at DuPont, and has since had various management positions.

Other Activities

When Bob Parry and I were young professors, aggressive and interested in our research, we attended the GRC regularly because we liked its format of talks in the mornings and evenings with afternoons free. I would play golf each afternoon, and tennis in the morning before breakfast. My golf game was fair, which allowed me to play with good golfers without interfering with their game. However, my golf never improved much, but I enjoyed playing the Laconia course because of the beautiful views surrounding it. As I got older, the younger golfers liked to invite me to play in their foursome, knowing they could beat me. However, at times when I played at my best, and when one or two in the foursome were having a bad day, I would win. My tennis, on the other hand, is very bad, but I was OK providing I got Denny Elliot as my partner. If he was playing well, we could almost beat Parry and his wife Marj.

At the beginning, the GRC had no money to assist with travel expenses for desired speakers from abroad. In spite of this, some invited foreign scientists would get their own funds and come. A few years later, the inorganic conference was able to get some funding from the Air Force Office of Scientific Research (AFOSR). Soon thereafter, the GRC itself began to make available ever increasing small amounts of money to each of the chairpersons as discretionary funds. The inorganic conference decided that these funds should be used to help foreign guest speakers pay travel expenses, and also to make some contribution to the expenses of graduate students and postdoctorates. The AFOSR finally had to terminate its support, but the GRC support has increased each year so it continues to help pay the expenses of some foreign guest speakers.

One of the early conferences had in attendance three scientists who were later to obtain the Nobel Prize: Ernst Otto Fischer, Geoffrey Wilkinson (1973), and Henry Taube (1983). Fischer and Wilkinson shared the Nobel Prize for their independent research on the sandwich compounds. At the time, Fischer was at the Technical University of Munich. Wilkinson had been a nontenured faculty member at Harvard and, as he did not get tenure, he had moved to Imperial College in London. I understand that when he got the Nobel Prize, he sent a case of champagne to Harvard, reminding them that they would not promote him to a tenured position. Henry Taube received the Nobel Prize in 1983 for his work on the mechanisms of electron transfer reactions of metal complexes. Taube was professor at the University of Chicago where he did his beautifully designed experiments using the reactions between Co(III) and Cr(II) complexes to illustrate examples of "inner-sphere" and "outer-sphere" mechanisms (pg. 86). Henry later moved to Stanford University. If I recall correctly, at that same conference there was also a graduate student named F. Albert Cotton, who has since become a famous inorganic chemist.[*]

The reason I have devoted all this time to the GRC is that I have been involved with it since the beginning of the inorganic conference. I have managed to get to this

---

[*]Having mentioned the GRC with three potential Nobel Prize winners, it is of interest to look at the participants of the Conseil de Physique Solvay. This conference, held in Brussels in 1911, must hold the record for having "all" the giants of science in attendance (**Fig. 4-11**).

conference almost every year, even now as an Emeritus Professor in my decrepit health condition. As I get older, many of the younger participants stay about the same age because the conference attracts new young inorganic chemists each year. Since I can no longer keep up with the voluminous amount of publications, attending the GRC gives me an easy way to keep in touch with the "hot" areas of inorganic research and with the young "stars" doing the work. I usually sit at the back of the lecture room and try to ask a question or two during the discussion. My questions are either of the pedestrian type to promote a little laughter or are on a bit of history to remind others of the adage that "there is nothing new under the sun."

My official role with the conference started years ago with an appointment on the Selection and Scheduling Committee. The duty of this committee is to select new conferences from among those suggested by any scientist and to monitor the conferences in an effort to know if they meet the standard required or if they should be terminated. My

CONSEIL DE PHYSIQUE SOLVAY
BRUXELLES 1911, HOTEL METROPOLE

1. GOLDSCHMIDT
2. NERNST
3. PLANCK
4. BRILLOUIN
5. RUBENS
6. SOLVAY
7. SOMMERFELD
8. LNDEMANN
9. DE BROGLIE
10. LORENTZ
11. WARBURG
12. KNUDSEN
13. HASENOHRL
14. HOSTELET
15. HERZEN
16. PERRIN
17. JEANS
18. WREN
19. RUTHERFORD
20. MME. CURIE
21. KAMERLINGHONNES
22. POINCARE
23. EINSTEIN
24. LANGEVIN

Figure 4-11. Attendants at one of the traditional Conseil de Physique Solvay Conferences held in Brussels, in 1911. There is a large framed portrait of the group at the registration desk of the Hotel Metropole in Brussels.

next official role was the result of being voted a member of the council. The council primarily keeps an eye on the conference status, trying to anticipate the needs of the conferences as well as their scientific direction in future years. My last tour of duty with the GRC was as a member of the Board of Trustees, serving as Chairman of the Board in 1975-76. The duties of the Board were similar to that of other Boards: keeping watch and making final decisions on all matters of importance to the GRC, such as locations of new sites, salaries of the director and other personnel, promotions, distribution of funds among individual conferences, etc.

During the years leading up to the Board appointment, I really got to know Alex Cruickshank and the two of us became the best of friends. Alex was a very frugal Scotsman who would not spend a dime of the conference's money if he could avoid doing it. He was, and is, a delightful, credible, hard-working person who one would like as a friend. As director of the GRC for 25 years (1968-93), Alex had many stories to tell about things done by conferees during his watch. For example, years ago the conferences in New Hampshire were held in a private boy's school where the bunks were not at all comfortable, there were no private washrooms, and the telephones were not in the rooms, but in the halls. Two of the earliest conferences were on petroleum chemistry and on catalysis. During one of these conferences, in the wee hours of the morning after BB (bring your own bottle) drinking, some of the members of the party decided to pull a prank on one of the most important not for fun attendees. Their leader knocked on the person's door and said he had an urgent telephone call. He immediately got up and, with pajamas on, ran to the phone. In the meantime the prankster shut the door of his room, knowing the fellow would be locked out of his room unless he had taken his key with him. There was this formal straight individual in his pajamas with no way to get back in his room for, at that hour, the GRC staff was not around. He had to awaken one of his fellow colleagues to get the emergency phone number of security and wait until they arrived to get back in his room. Needless to say, he was able to find out who the culprit was, and he saw to it that this person was never again allowed to attend the research conference on catalysis. This prank also resulted in the door locks being changed so they could only be locked by someone inside the room.

Finally, one more item of interest to me—I mentioned how frugal Alex was, in particular with GRC money. Since he felt friendly and comfortable toward Denny Elliot (AFOSR), Robert Parry (U. Utah), and me, he would often want to talk with us former Board members about some of the conference's problems. At one point one of us said, "Alex, as we are talking about the conference's business, we could just as easily talk about it over a dry martini and a couple of lobsters. This would even justify your having the conference pay the bill as a business expense." This is how our Tuesday inorganic GRC lobster lunch began, which since has become a tradition, now continued by the current Director Carl Storm. The "lobster" group has grown to include three other former chairs of the Board: John Fackler, Alan Cowley, and Harry Gray.

From this long write up it should be clear that the GRC has been of great service to inorganic chemistry, other areas of chemistry, and to science in general.

INTERNATIONAL CONFERENCE ON COORDINATION CHEMISTRY (ICCC)

The ICCC, started by Dr. Joseph Chatt (later Professor at the University of Sussex), was first held at Welwyn, United Kingdom, in 1950. Joe Chatt was an outstanding coordination chemist with the vision that this area of chemistry was in its infancy with the potential to grow and become a major area of chemistry. That he was correct is seen by the present intensive research and scientific publications involving the coordination of ligands to metals in some capacity. For his seminal important research on these types of compounds he received many awards and the one I recall best is the Wolf Prize, which is considered second to the Nobel Prize.

Chatt invited a small group of about 30 inorganic chemists to participate in an initial meeting, later to become the ICCC. Most of the attendees were English, Scandinavians, or Germans doing research involving coordination chemistry. No Americans were present. The meeting was such a success that Professor K. A. Jensen at the University of Copenhagen organized the next meeting in Copenhagen. This gave birth to the ICCC which is now held every other year, except for additions, at times, in some distant country less accessible to possible attendees. The initial conferences were held in Amsterdam, Rome, London, and Detroit. A total of 33 conferences have been held in different countries worldwide. This year, 2000, it was held in Edinburgh and honored Joseph Chatt. Of all the ICCCs, this is one that I surely would have wanted to attend. However, because of my health, I could not travel that distance. Joseph, his family, and mine have been good friends for the past 45 years. He and I exchanged Christmas letters each year and his were always twice as long as mine. I kept in touch with his chemistry and have always admired his important research which was done with the greatest of care. We have one joint publication with Professors Ralph Pearson, Harry Gray, and Bernard Shaw (pg. 99). In memory of Chatt and his chemistry, the Royal Chemistry Society of London has established an annual award of the Chatt Medal. I am very proud, but humble, to have been the first Awardee of the Chatt Medal.

In the year 2000 the 34th ICCC was attended by more than 1,500 participants. It will have seven plenary speakers and will cover the following six topics: Structure and Dynamics, Biotechnology and Medicine, 21st Century Materials, Technological Advances, Chemistry of Life, and Joe Chatt Chemistry.

My own involvement with the ICCC stems from my now being one of the oldest chemists doing research on coordination chemistry, having started with my dissertation research in 1940 with my mentor, Professor Bailar. My academic research has always been on fundamental problems involving coordination chemistry.

The first ICCC I attended was its third conference, held in Amsterdam in 1955. I always remember this conference because Professor Walter Hieber's answer to my question after his lecture was largely responsible for our starting our research on metal carbonyls (pg. 101). The other reason I remember that conference is that I met Professor E. O. Fischer for the first time. He and I became very good friends, and even had a joint program of exchanging predoctoral graduate students in their last year of research.

During this conference E. O. and I were staying at the same pension. One morning we arrived for breakfast at the same time, introduced ourselves, and had breakfast together. This was only the second day of the conference, but he said "We will meet again, for I must return home this morning." I asked, "Why before the conference is finished?" His reply was, "I must make certain people in my research lab keep working very hard, because the Harvard group is always working very rapidly." He was referring to research being done by Geoffrey Wilkinson. The two groups were working diligently to prepare as many as possible of the newly discovered cyclopentadienyl metal compounds known as "sandwich" complexes. Wilkinson and Fischer carried on a bitter polemic in their publications, but in 1973 they both were willing to share the Nobel Prize. Peter Pauson, who discovered sandwich compounds with his synthesis of the *bis*-cyclopentadienyl iron compound or ferrocene, did not get included to share the Nobel Prize. He had made this compound when he was in Duquesne University, where he formulated it incorrectly and before he had the opportunity to do further research and correct his postulated bonding and structure. He could not compete with the large capable research groups of E. O. and Geoffery that were able to investigate ferrocene in detail and to realize that most metals could make similar cyclopentadienyl compounds.

Figure 4-12. Professor Stanley Kirschner of Wayne State University in Detroit, Michigan, who, as Secretary of the ICCC, kept it alive and improving from 1961 through 1989 at Brisbane, Australia. To Stan's left is his able successor, Professor Jan Reedijk of Leiden University, the Netherlands.

During the intervening years, I have been involved with most aspects of the ICCC. This means I have been a plenary lecturer a few times, including the recent (1998) conference in Florence. When invited to give this talk in Florence, I had to tell them I had no new chemistry to present, because I had not even had a lab for the last eight years. They were aware of this, and asked that I give the talk I have given many times on the early history of metal complexes. It was felt that most of the 1,000 or more coordination chemists who would be present knew little about the early history of these compounds. The talk was well received and some said it was much better than had I talked on our research. My other involvements with the ICCC include giving shorter talks over the years on some specific aspects of our research, chairing different sections of research, and, when attending a meeting, being a representative of the US in policy matters and the selection of future sites for ICCCs.

Finally, I was approached to hold the first ICCC in the US at NU in 1961. I had to refuse, and this was fortunate because Professor Stanley Kirschner (**Fig. 4-12**) organized an outstanding conference at Wayne State University in Detroit. Even more important, he was appointed Permanent Secretary of the ICCC and served superbly until he resigned in 1989 at the Brisbane, Australia Conference. Stan did an outstanding job as secretary and has to be acknowledged as having been a prime factor in helping the ICCC become what it is today. He was succeeded by Professor Jan Reedijk (**Fig. 4-12**) at Leiden University in the Netherlands, who has got off to a running start, assuring us that the ICCC is in good hands. The second US conference was hosted by Professor Robert Sievers at the University of Colorado at Boulder in 1984. Not previously mentioned is the special evening outing at each conference, which usually involves attendance at a symphony, opera, or play. The Boulder Conference was unique because Bob arranged to have the group attend a real live rodeo.

The ICCC conferences are among the best of the conferences which I have attended. They always report good, current chemistry, and provide an opportunity to meet many foreign chemists. The ICCCs have played an important role in my professional life, for which I am very thankful.

FUNDING AGENCIES

Most academic research-oriented chemists with agency funding will, at some point, be asked to serve on a panel to review research proposals and recommend those which are worthy of financial support. In the US, some of these agencies include PRF (Petroleum Research Foundation), NSF (National Science Foundation), NIH (National Institutes of Health), DOE (Department of Energy), and NATO (North Atlantic Treaty Organization). I therefore have been asked to serve on such committees and, fortunately, most have an upper limit of five years of service, with another five year renewal possible. Four of the panels that I served on were PRF, NATO, NSF, and NIH.

The PRF Advisory Board

The Petroleum Research Fund is a trust fund which was established in 1944 by Standard Oil Company Incorporated and several other oil companies. PRF was to be a charitable, scientific, and educational trust, with the ACS named as the qualified recipient of the income therefrom. The ACS receives an annual income from the trust, allowing it to provide support for scientists in different categories. In 1998 the PRF provided new grant support as follows: Fundamental Research in the Petroleum Field (183), faculty in non-Ph.D. granting departments (48), starter grants (30), and scientific education (50). New grants, plus the on-going grants, are now being supported at a total amount in excess of four hundred million dollars.

I was a member of the PRF Advisory Board for five years (1968-73). The Board members met three times a year, twice at the ACS building in Washington, DC, and once at some other location. During the period when I served on the PRF panel, one of the innovations attempted was to select an outstanding person doing research on petroleum or its related chemistry. This person would receive an Award of $100,000 to support his research. This was a lot of money then. This award did not require a submission of a research proposal. Since the Award gave the panel an opportunity to pick a high quality person, it likewise would stress the mission of the PRF—petroleum. Unfortunately this program was short lived because a few chemists expressed some concern. One even went so far as to write a letter to the panel stating that his research was much better than that of the first Awardee and he should have been picked for the Award. The Award was discontinued after a couple of years.

As is true of all panels, its members read and evaluated the candidate's research proposal. Next the responses of 4-6 outside reviewers, selected because of being knowledgeable in the area of work proposed, are discussed and given much weight. The panel members thoroughly review all of the available information for each of the proposals, making certain that all information available is being fairly considered. Subsequently, three lists are prepared: (1) proposals that should be funded, (2) in order of preference, proposals that could be funded if there were sufficient funds, and (3) proposals that should not be funded. This is always an advisory report to the agency, but it is rarely changed.

Figure 4-13. Professor Melvin S. Newman (1908-93), at Ohio State University, is well known for his research on reaction mechanisms of chiral compounds. He introduced the Newman Projection for representing the geometry of molecules, and it is used in organic textbooks.

Another member of the PRF panel, Professor Melvin S. Newman (**Fig. 4-13**) of Ohio State University in Columbus, Ohio, and I both liked to play golf, so we planned our schedules to include golf before and after all of the meetings that were held outside of Washington, DC. We would arrange our flights to arrive early enough the day before the meeting to play golf that afternoon. The day after the meeting we played another 18 holes of golf in the morning and returned home in the afternoon. Two of the courses I recall best were the Chi Chi Rodriguez Course in Puerto Rico and Torrey Pines in San Diego, California. The Chi Chi course I remember because Chi Chi was an outstanding professional golfer and remains so in the senior tournaments. He is always kind, colorful, and charismatic, even when playing in a golf tournament, so many of the spectators in the crowd like to watch him. Furthermore, the Chi Chi course is near the ocean with two golf holes providing a beautiful view of the blue Atlantic Ocean. The other course we enjoyed playing was Torrey Pines. This too is a course with some of its holes overlooking an ocean—this time the beautiful blue of the Pacific Ocean. If lucky, one might even see whales swimming. This course also gives a good view of the Salk Institute, named for Dr. Jonas Salk, who discovered the vaccine for polio. The last time

I saw Mel was in 1992 when I was a Distinguished Visiting Professor at Ohio State University for one semester. I went to his office to visit him, only to find him in his lab doing hands-on research. He was standing on a stool reaching for some glassware on a top shelf. This can be dangerous, and we always caution students not to stand on a stool in the lab, yet Mel was doing it, even at his age. With all of this going for Mel and me, it is easy to see why I enjoyed my tour of duty on the PRF.

North Atlantic Treaty Organization (NATO)

NATO is well known as a political/military organization with 15 members (Belgium, Canada, Denmark, France, Germany, Greece, Iceland, Italy, Luxemburg, Netherlands, Norway, Portugal, Turkey, United Kingdom, and the United States). How does it happen that NATO has had a basic science program the past 42 years? It was noted, after a more careful reading of Article II of the North Atlantic Treaty, that its conditions allowed for basic research to help insure the "well-being" of the NATO member countries. This was discussed in 1956, with the goal of preparing a case for the need of a program on basic science. This caused some unease among some members of the alliance. However, the launch of the first sputnik by the Russians in 1957 helped to bring in focus the need for such a program. In 1958, the science committee of the basic science program had its first meeting.

In 1971 I was invited to serve on the panel of the NATO Research Grant Program. Each NATO country had its own representative on the panel and I was to be the US representative. The committee met three times a year (in February, July, and October), twice at the NATO headquarters in Brussels, and once in a "science developing" member country. I replaced Taube, who had finished his five years on the panel and had suggested me as his replacement. This was a rather unique panel because it had to deal with proposals from all areas of science, engineering, and mathematics. The panel was chosen so as to have at least one expert in each of these areas of science. Almost all of us were Professors, and it was interesting to work with mathematicians, biologists, etc. and to see how they handled research proposals in their fields. (American scientists will be shocked to know that instead of the many paged research proposals required in the US, only a one or two page proposal is needed in Europe.) Decisions were largely made on the basis of the "track record" (number and quality of scientific publications) of the principal investigator. Young investigators were also judged on their track records, with more emphasis placed on the quality of their publications, rather than the quantity. When possible, weight was also given to the promise of success of the younger investigator. Needless to say, this was a learning experience for me, since all research proposal panels for chemistry in the US are composed only of chemists.

I was Chairman of this panel during my fifth year, and was then asked to stay an additional year as Chairman. I was about to say no, because traveling to Europe three times a year for five years was a bit much, but I finally stayed the sixth year. The developing science countries we visited during my six years on the panel were Turkey, Italy (Sardinia and Sicily), Portugal, Greece, and Iceland. An attempt was made to help the science in these countries. For example, all NATO funding for this program involves the collaboration between two or more investigators in two or more NATO countries. The panel would always try to be more supportive if one of the countries was a science developing country. It was understood that a principle investigator in a science developed country should try to find someone with whom to collaborate in a science developing country. In each of these countries, we received a warm welcome and special treatment, but I think our most interesting visit was to Iceland. This was a fascinating country for me to visit. It has a rugged terrain, much of it covered with what, at some time, was volcanic lava. When driving to the airport, we learned that one must avoid going over some of the sharp edged rocks which are hard on the car's tires. I could not believe that persons were swimming in a large outdoor swimming pool, when we were cold wearing heavy winter clothes. Later we were told the water's temperature was kept warm by the use of volcanic thermal heat. We also were shown large gardens with the usual vegetables, including nice red tomatoes. The enclosure for the vegetables were volcanically heated. It was exciting to see how it was possible to bring this subterrainian heat to the surface and use it for a variety of purposes.

Not long before we visited Iceland, a large volcano erupted on the small fisherman's island known as Westman Island. The island had about 4,000 inhabitants, and their one industry was commercial fishing. This island was suitable for fishing because it had a natural harbor for boats. The eruption of the volcano came after a three day warning that it would arrive, which gave the people time to leave their island and, thus, there was no loss of life. Our panel members were flown to the island with a university geologist as our guide. What we saw was a totally destroyed island with all of the houses damaged beyond repair, and some literally buried under volcanic ash. The geologist told us that the lava flow, which moves slowly, was going in the direction of the fishermen's harbor. If the harbor was to fill with solid lava, it would be rendered useless to the fishermen. In order to prevent this, a desperate call was made to all vessels on the North Sea near the island to come and help. Strong forces of large amounts of water were directed to the front end of the slow moving lava which caused the lava (molten iron) to solidify and prevented the stream from continuing on its direct course. In fact, the lava was finally directed to fall along the side of the harbor, thus even increasing its size to accommodate more fishing boats. The geologist asked us to touch the surface of the soil we were standing on, and it was still hot to the touch. The only other persons on the island except us were drivers of the bulldozers working to prepare the island for its people to return.

NATO always reimbursed us for first class travel. When I asked if I could use that amount for two coach tickets instead, thus giving me a chance to take one of my four children with me, the answer was yes. This made my children very happy and I took them with me in terms of their age. This meant that MC was the first to go. We took advantage of the lower air fare for trips of 14 days. This gave us more than ample time to go sightseeing after my three days of panel meetings in Brussels. While there, we were put up in the Metropole Hotel where a group of the giants of science stayed many years ago when attending a Solvay Conference (see **Fig. 4-11**). We were also pleased when we found out that NATO had paid our hotel bill. During the remainder of our travels we found less elegant, but also less expensive, hotels. I would rent a car and we would drive around to see a couple of the small countries nearby. Most of our final week was spent in London at a bed and breakfast hotel in an Italian area near University College London, where I had stayed several times before. We had plenty of time to see almost all of the tourist sights and to attend a few of the plays. Much of the same schedule was used for the other three children. However, one difference that took place when I traveled with Fred was that we stopped a couple of times to rent clubs and play golf. One golf course that interested us very much was in Waterloo, the town of Napoleon's last stand.

My trip with Liz resulted in something I did not believe possible. Since the stay in London was longer than necessary to see "everything," we spent the last day or two waiting for our 14 day stay to be over. One day Liz and I watched the Wimbledon tournament on TV. The following day was to be the ladies final— pitting Billy Jean King against Chris Everett. Liz, at age 13, had attended a tennis camp and was very involved in tennis. She knew the names of the top women professional players, and she desperately wanted to attend the final championship match. I tried to explain to her that this was not possible because, like the World Series or Super Bowl at home, all of the tickets were sold. She would not take no for an answer, so I took her there to convince her that it was impossible to get tickets. Instead, scalpers, seeing that we were not English, decided they could get more money from tourists and descended upon us. They charged more than I expected, but I gladly paid it so that Liz could have her once-in-a-lifetime chance to see the ladies final match of the Wimbledon tournament. We had the cheapest tickets, to be used for standing room. As it turned out, this was a good place to be, because we were close enough to the court to hear how hard the balls were hit, how fast the balls traveled, and even the grunts of the players. For Liz, it meant she was close enough to take pictures of the players. She took many pictures only to find out when she got home and had films developed that the players were too far away—they could be seen but not identified. Nevertheless, Liz was very

Figure 4-14. Professor Renato Ugo, at the University of Milan, is known for his early research and book on the similarity and differences between heterogeneous and homogeneous catalysts. In addition to his academic career, he also spent some years as a top executive in the large chemical company Montedison.

pleased to show her pictures to her tennis friends, and to tell them about the tournament.

One final assignment that I had from the NATO Research Grants Program was to arrange a conference involving two or more different, but related, areas of science research. This assignment started when I received a telephone call from the Deputy Secretary General for Scientific Affairs. He asked if I could come immediately to his office to discuss the possibility of some such conference. I said yes, and the next day I was on an airplane on my way to NATO headquarters in Brussels. There I was asked if I knew an area of science where the very top investigators differed in their views and research approach to the same problem. Fortunately, I was chairman of our department at the time and I had just been involved in preparing a research proposal on catalysis. The proposal amounted to a three-way attack on catalysis. The three ways were heterogeneous, homogeneous, and metalloenzyme. They listened to what I had to say about this, then enthusiastically said this was exactly what they had in mind. They told me that they had to move quickly for they were about to end the year with a $50,000 surplus. This they did not want to return, because each year they asked for an increase in their budget so they wanted to make sure all of their money was spent.

Figure 4-15. Professor Robert Burwell, Ipatieff Emeritus Professor at NU. He is mostly responsible for modifying surfaces on solid supports to produce a good catalyst for certain processes. Bob was the most knowledgeable scholar on our faculty. Regardless of what the topic of discussion was at our brown bag lunches, he always provided us with a clear explanation—even if, at times, we were not certain if he was correct.

I went into immediate action. I first called Joseph Chatt, an expert on nitrogen fixation (metalloenzyme) and then Renato Ugo (heterogeneous, **Fig. 4-14**)–I would handle the homogeneous–and told them the reason for the urgency and asked that they come at once so that the three of us could work on the program as well as decide whom to invite. Since this was the end of November, we had much to do before sending out invitations early in December. We needed a secluded quiet place where we could all be housed and get our meals. Ugo was able to get us the ideal place for our workshop. This was at Santa Margherita di Pula in Sardinia. It is near the coast, and is used by Italians for their summer vacations. In the winter, it is used only for special groups such as ours.

During the workshop, no one was to leave the premises and the workshop would make certain everyone contributed to the work load. I asked my colleague Professor Robert Burwell (our expert on heterogeneous catalysis, **Fig. 4-15**) to help me, and together we made certain each group gave us their report before leaving. This was before xerox copies, which meant using the old mimeograph to make copies. We had to get all of their reports before leaving to return home because if we asked them to send us their reports, it would take "forever" to get the desired book in print. It is amazing that we were successful in getting all of these busy experts to come on such short notice, and just a week before Christmas. This shows that, with all expenses paid for a week in a very fine place, even the best of the bests can be attracted if a program is first rate. We were pleased that we were able to attract a real "scientific dream team" (see Appendix). The book, *Catalysis, Progress in Research*, was published by Plenum Press with my colleague Bob Burwell and me as editors. The book not only summarizes what is known about catalysis, but, even more important, what is not known and needs further study. I have been told by different young chemists writing research proposals for funding that they found the book useful in giving them ideas for research. Hearing this has been very satisfying, for it was exactly what we had in mind.

5

COUNTRIES AND CHEMISTS VISITED

This chapter is devoted to a few of the countries and chemists that I have visited. In preparing for this chapter, I am surprised to find that I have been to as many as 41 countries (see Appendix). This may be why some say that "a full professor is one that is gone full time."

The countries were visited in response to invitations to give lectures at universities or international conferences and to represent the ACS when President in 1983. Chemists of my generation who watched a TV show called "Have Gun, Will Travel" may, at times, say "Have Slides, Will Travel" because many of us travel worldwide with our research slides. These trips, although numerous, are taken mainly during the summer or in the quarter (or semester) when one is not teaching. There are two reasons for accepting all these invitations: (1) they perform a service to chemistry and (2) they focus attention on one's university and, particularly, its chemistry department. Be that as it may, I do not plan to burden the reader with a discourse on the 41 countries, but will select a few which were of most interest to me.

ITALY

I have been to and from Italy more times than I can recall. To me, Italy is my second country after the US. This is surely because my parents emigrated from there and I know a little of the language. As indicated in the first chapter, I cannot recall my initial trip because I was six months old when our family went there, intending to stay but did not, and only a year old when we returned. What I consider to be my first time in Italy was in 1955, when Mary and I left our two children in Copenhagen and drove through parts of Europe on our way to Rome (see *Chapter 2*, pp. 37-60).

My next trip to Italy was to attend the ICCC in Rome in 1957. I presented a talk on our research on the mechanism of platinum (Pt^{2+}) catalysis of the ligand substitution reactions of Pt^{4+} complexes. I went to hear most of the talks presented and to some of the sightseeing tours arranged by the conference. I met many coordination chemists who knew me, and whom I knew, through our publications. It was a useful, but uneventful, week for I had to return directly home after the conference.

My next trip to Italy in 1959 resulted from an invitation by my friend Professor Luigi Sacconi. He had recently been promoted to Professor at the University of Palermo. The promotion pleased him, but being a staunch Florentine, he was not pleased with having to start his career in Palermo. However, this was the Italian way of being promoted to Professor, first being appointed to a smaller university before being appointed to a larger one.* Professor Sacconi was later transferred to Florence, where he spent his entire professional life. When he invited me to come to Palermo for a few days to lecture, I told him that I was teaching, but would let him know when I could come. A few weeks later, I received a first-class air fare ticket to Palermo. A letter stating that they were expecting me in Italy and an itinerary for my two weeks in Italy were also enclosed with the ticket. I was about to return the ticket and remind him that I had said I was teaching a course and would have to come at a later date. Then I recalled being told that Luigi would never accept no as an answer. I asked advice from my colleague Selwood, and he surprised me by saying that I should go and that he would teach my class for the two weeks. My itinerary was to spend four days in Palermo, two days in Rome, two days in Florence, and two days in Milan.

Sacconi had moved his assistants, graduate students, and lab facilities from Florence to Palermo. I arrived prepared with many slides, knowing that, at the time, rather few Italians knew English. I gave my first lecture in English, but could see that the attendees, numbering about 25, were having a difficult time following my English. After my talk, there was little or no discussion. That evening at dinner, I said to Luigi, who spoke very little English, that perhaps I should try to give my second lecture in Italian. I told him that since my parents were not chemists, I only could use my peasant dialect, Piomentese, for ordinary discussion, and would use English for the chemical terms. Before my second lecture, Sacconi told the group that I would try to give my talk in Italian. The same group that had not seemed turned on during my first lecture were wide awake, at times whispering and smiling to each other, because I spoke very poor

---

*University faculty in Italy were promoted to associate professor by a committee of senior professors (**Fig. 5-1**) who examined their record as assistant and conducted an oral defense of the record by the candidate. This is called Concorso, and as a result of this examination, the persons who pass are promoted to the rank of professor, and assigned to one of the smaller schools. When successful there, the professor is transferred to a larger, more important university. I am told that often the senior professors make deals with one another to support each other's candidate for professorships.

Figure 5-1. The senior inorganic and analytical chemistry professors who met in Rome in 1962 to conduct a "concorso" (oral defense by the candidates) in order to select young faculty for promotion to rank of professors. My host, Professor Caglioti, being the most senior member of the group, is number four from the left, near the front.

Italian. At the end of the loud applause, there were several questions and discussion of the chemistry presented in the lecture. Luigi and I agreed that the "Italian talk" was better than the first given in English. Likewise, a few members of the group told Sacconi and me that they enthusiastically preferred the second lecture and hoped that the final one would be the same—which it was.

My next stop was in Rome for two days, where I was hosted by Professor Vincenzo Caglioti who, as the most influential inorganic chemist in Italy, had a large research group. He said to me, in Italian, for he did not know English, "I hear you gave your lectures in Italian. This is not Palermo, so you can give your talk in English. I have a large research group and some know English, having spent time in England." I told him I would gladly give my lecture in English, but that my talk would be presented in three related parts of our research. I suggested that I give the first part in English, the second in Italian, and then ask for a show of hands as to which they preferred for the final lecture. The show of hands for the third part to be in Italian was unanimous, with Caglioti standing up in the front row, waving both hands. After that, my remaining talks in Florence and in Milan were given in Italian.

In Milan I gave my lecture at the University of Milan where my host was Lamberto Malatesta. Present at my lecture was Professor Guilio Natta (1907-73; Nobel Prize, 1963) and some of the members of his research group at the Istituto di Chimica Industriale del Politecnico di Milano. His research group was one of the very best in organometallic chemistry in the world. After my lecture he invited me and some of his assistants to his house for dinner. We had an excellent dinner, but the thing Natta (**Fig. 5-2**) wanted us to notice was his new suit made of isotactic polypropylene, which he and his group discovered and which would get him a share of the Nobel Prize with K. Ziegler in 1963. One other item he showed us was a Scottish colored blanket made of the polymer, which demonstrated that the colorless polymer could be colored by nontraditional means.

I believe that some of those who heard my talks are still laughing, as they tell me how much they enjoyed my talk for the good chemistry, but even more so, for the bad, but amusing, Italian. I suppose, since I was able to make such a fool of myself, the Italians adopted me as "Basolo is one of us and he must be a good guy." In any case, the Italian chemists have been very generous to me. In 1981, I was made an Honorary Member of the Italian Chemical Society at their annual meeting which, that year, was held in Catania in the southeast of Sicily. The attendees were all Italians, except for a very few foreigners like myself. The Italians presented their research in Italian. I was able to further develop my Italian listening skills, understanding about 25% of what was being said.

I became aware of the relaxed nature of the Italians one day at lunch. There was a talk I wanted to hear at the beginning of the afternoon program. The Italians at my table were finishing their meal and wine. I announced that I wanted to hear the first talk after lunch and would have to leave. They said, "You have plenty of time, because the speaker is still here at the next table." When the speaker left, I followed, and I heard the entire talk, which was given some 15 minutes late.

During the meeting, some of us went to see Mount Etna which, at the time, was on its good behavior. I quickly learned that the Italians are aggressive drivers, and this means a foreigner should spend a few days driving carefully to become accustomed to how Italians drive. After that, most foreigners learn to drive as aggressively as the Italians. However, even the Italians say that Sicilian drivers are dangerous, and the worst ones are in Catania. I can confirm this by a couple of things I saw when I was there. Some of the drivers looked from side to side when arriving at a

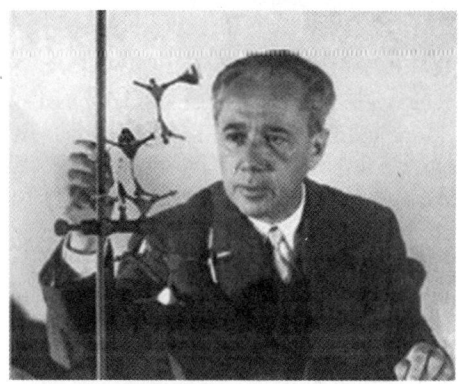

Figure 5-2. Professor Giulio Natta (1903-73; Nobel Prize, 1963) was the chairman of the Chemical Engineering Department (Politecnico) at the University of Milan. He surrounded himself with an outstanding group of organometallic chemists. He is shown here viewing a model of his isotactic polypropyline which earned him a share of the Nobel Prize with K. Ziegler in 1963 "for their discoveries in the field of the chemistry and technology of high polymers."

red light and, if no one was coming, they drove on through the intersection. I saw this happen when a police car was parked nearby, but the police did nothing. In fact, the Italians tell the story about a driver who was brought to court because he drove into the back of a lady's car, doing considerable damage. His defense was that, although the street light was red, there were no cars coming, and he did not expect her to make a sudden stop. Therefore, he had *increased* his speed to also make it through the light. The case was dropped, and the insurance paid for repairs to the car.

Figure 5-3. A photograph of the President of the Italian Academy of Science Lincei presenting me their certificate of foreign membership to the Academy in 1987.

Figure 5-4. Professor Luigi Sacconi, to the left, and Lamberto Malatesta, to the right, of me after my having become a foreign member of their Academy, Lincei. We thank Professor Ivano Bertini of the University of Florence for taking the picture.

One other thing that people who have spent time with Italians know is that they "always" use their hands when they talk, particularly if they get excited. This has always been done, even when they talk over a telephone, and, now, with the cell phone, I have even experienced someone driving on the interstate with no hands on the steering wheel, albeit for only a short distance. I was sitting in the front seat next to the driver when he got a call on his cellular phone. He talked for some time to the person who called, but at some point he got angry and made gestures with his hands off the steering wheel. I told him to please keep his hands on the steering wheel, for although I liked Italy, I did not want to die in Italy.

The highest honor accorded me in Italy was to elect me to their foreign membership of the *Accademia Nazionale dei Lincei* in 1987 (**Fig. 5-3**). This is Italy's National Academy of Science, having only 10 foreign members—five of whom are Nobel Prize laureates. After I had gone through the ceremony to sign my name on their document and become a member, Professor Malatesta congratulated me and said, "Now that we have elected you a member, we expect you to get the Nobel Prize, as have others whom we have elected as members." I told him I would gladly accept the Prize, but it was up to them to nominate me and get others to do the same. Malatesta, Sacconi, and I then went to have a wonderful Italian lunch (**Fig. 5-4**).

This Academy has to be one of the oldest and most prestigious. It was founded in 1603 by four young noblemen. However, the Academy took on the importance it now has starting with Galileo, who became a member in 1611. The name Lincei was chosen because the lynx can see in the dark, and members of the Academy are those with the intelligence to see what others cannot see.

There have been several other honors (two Honorary Degrees and three Medals), but I will only mention two here because the others are given in the Appendix.* I have many chemist friends in Italy, mostly faculty at universities, but also others in the

Figure 5-5. A picture of me receiving the *Laurea Honoris Causa* (Honorary Degree) from the University of Turin in 1988.

---

*One of the Honorary Degrees I wish to mention is that from the University of Turin (**Fig. 5-5**). Turin is in the Piedmont region from which my parents emigrated (see Chapter 1). Therefore, to be so honored near where my parents had lived was particularly important to me. It was also important to receive the *Laurea Honoris Causa* in Turin because I still have distant relatives who live nearby and were able to attend the ceremony, and with whom I visit each time I go to Italy. My other *Laurea Honoris Causa* is from the University of Palermo.

Consiglio Nazionale delle Ricerche (CNR) labs or in industry. I appreciate all they have done for me over the years, and I wish them all continued success in their personal family lives and in their chemical research and teaching.

I also must relate one final point about my many visits with my cousin Savina Sgarbi and her family in Milan. On each of my trips to Italy, I would fly from Chicago to Milan where I was met by a member of the Sgarbi family (**Fig. 5-6**). I would be a guest at their house for a couple of days, then go to where I was to attend some scientific meeting. Before leaving Italy, I would again spend a day with the Sgarbi family. They are a superb family, and I watched the three small boys grow up to become responsible young men.

Figure 5-6. My relatives, the Sgarbi family, except for their oldest son. Savina is at the table in front of her husband Romano, Fabbrizio is at the table and Dario is in the back near his father. They are not only distant relatives, but they are very good friends, and I love all of them.

PEOPLE'S REPUBLIC OF CHINA

I have been to China seven different times, and here I will give an account of some of my visits there. My visits to China started with a telephone call in 1951 from Professor Nebergall of Indiana University. He said that his Chinese graduate student, Yun-Ti Chen, had just defended his Ph.D. dissertation, and would like to come and do research in my lab. However, he would need some financial support and I had to tell him that I had only $2,000 from the Research Cooperation, which would not be enough to take care of him for a year. In spite of this, Chen arrived at the end of 1951, and managed to stay two years. He was my first postdoctorate, for they were not common in those days. At about this time, the US and China discontinued diplomatic relations. This made it impossible for Chen to get a visa to return to China. He had to go first to Europe, and from there to China. Shortly after his return, Chairman Mao started the terrible program called the "Cultural Revolution," that required all able-bodied people to work in the fields in an attempt to prevent poverty and starvation. Since professors were no exception, this included closing the universities. The people had to do hard work for long hours each day, making it particularly difficult for persons not accustomed to such farm work. Chen was one of these people, but finally, when this program came to an end, he returned to his home and to his faculty position at Nankai University in Tianjin.

After Chen left our lab and managed to get back to China, he did not write to me until after the US and PRC renewed diplomatic relations in 1974. Soon after the agreement between President Nixon and Chairman Mao, I received a letter from Yun-Ti Chen, Professor at Nankai University (**Fig. 5-7**). He explained that he was back to his faculty position at Nankai University, he was married, and they had two children. He knew that Pearson and I had written a book, but he could not get a copy. He asked if I would please send him one. I did, and he wrote to thank me. On a later trip to China, I visited their libraries and found our book in all of them. There were also the journals of the ACS, including *Chemical Abstracts*. Since China had no copyright agreements with the US, they would make as many copies of journals and books as needed.

In 1979, I received a letter from Chen saying that he had a program of the ICCC meeting in Calcutta, and saw that I was scheduled to give the opening plenary lecture. He said the Ministry of Education wished to invite me and my wife to be their guest in China for three months. During this time, I was to give lectures in a course on inorganic coordination chemistry. They could not pay travel expenses, but they said since our travel expenses to India were being paid by ICCC, we need only to stop in China on our way to India. They would provide meals and lodging for both of us in China. Mary and I discussed this, knowing that relatively few Westerners had gone to China prior to 1979. We did know that two NU faculty had been to China to give invited lectures. They both enthusiastically told us to go, but to be prepared for a cultural shock and for the primitive conditions of a developing country. We finally decided to accept the invitation if we could be there for only one month. This posed no problem, so Mary and I flew to Peking (now Beijing), stayed a month, and then went on to Calcutta.

Mary and I agreed that our stay in China was the most exciting and memorable country we had ever visited—and I still feel the same. We knew that there was no gender difference in their dress (**Fig. 5-8**), and this made it difficult, when at some distance, for us to distinguish between female and male. However, this was showing some signs of changing because we saw a few women wearing western-type dress. This was mostly seen in the city of Shanghai, and when I returned four years later, it was almost commonplace.

Figure 5-7. Yun-Ti Chen, Professor at Nankai University in Tianjin, PRC, and me. Yun-Ti was my first postdoctorate, arriving in 1951, when postdoctorate chemists were not common.

Our first three days we stayed in Peking, with a room in the Peking Hotel. This was said to be the best hotel in all of China. It was satisfactory, but not elegant. One morning I went out on the balcony of our room on the fifth floor, and looked out over the city, only to see a dark cloud of smoke from their burning of soft coal. Furthermore, the Chinese are heavy smokers, so that one often sees spittoons, even in the Hall of the People where President Nixon and Chairman Mao held their

Figure 5-8. A photo of the approximately 200 faculty from all areas of China who attended my one month of lectures on coordination chemistry at Nankai University in 1979. The Chinese Ministry of Education had invited me and my wife to come, and for me to give these lectures. They also encouraged Chinese chemists to attend, and paid their travel expenses. This shows that there was no gender difference in their dress.

meetings. One thing we found annoying at the hotel was that the toilet in the washroom kept making the sound of trickling water. Later, we found this to be true for all toilets. One evening, during a conversation about this problem, one of the Chinese said, "Yes, we can send a rocket to the moon, yet we cannot fix our leaking toilets."

One thing we were very pleased about is that neither their water nor their meals made us ill. This is surely because they boil their water and give it to you in large thermos containers. Their food is cooked at very high temperatures to kill any bacteria. No one should ever eat salads, or any other uncooked foods. I solved my drinking problem somewhat by drinking their very best beer instead of water. The beer (Qing-Dao) is exported and can be found, at times, in our larger liquor stores. The local beers are terrible and so is their hard liquor (Mao-tai, although I was told that President Nixon liked the liquor), as is their wine.

During our stay in Peking, the Ministry of Education arranged for us to have a car and driver at our disposal. Our host, Professor Yun-Ti Chen, kept us busy, wanting to show us all the tourist sights in and surrounding Peking. We started with the Great Wall, the Forbidden City, etc. I told Chen that, having grown up in the country, I would like to see the countryside. We were driven around a bit, but the roads were so bad that the ride was none too pleasant. We walked around Tienemin Square, which is a large spacious area in front of a large portrait of Chairman Mao and of the Forbidden City. One thing that troubled Mary and me were the large photos of Lenin and Stalin. I asked Chen, whom we knew well enough to ask, "Why is Stalin so largely portrayed when everyone knows that he had hundreds of thousands of his people in Russia massacred?" Chen said, "He is shown there because he is the one who industrialized Russia, making it a wealthy, powerful country." It must be realized that China was trying desperately to become a prominent, formidable, industrial country. While in Peking, every evening we were taken to an interesting performance, where knowing the language was not required. This meant shows of mostly songs and dance, magicians, and athletic performances. These were different from anything we had ever seen at home, and the persons involved were outstanding.

After our three days in Peking to rid us of the culture shock, our host took us on a train to Tianjin and Nankai University where I had to go to work giving my lectures. The Ministry of Education had it announced that I would be giving these lectures on inorganic coordination chemistry and, furthermore, that the participants' travel and living expenses for the month would be taken care of by the government. No less than 200 came by train from all parts of China. A few told us that it had taken them three days to get to Tianjin. Most of the persons attending knew little or no English. I had been warned that this would be the case, so I prepared many slides to clearly show the formulas of compounds being discussed, along with chemical equations and experimental graphs and data. They provided a translator, and I talked very slowly. A topic that I would cover in 50 minutes in the U.S. required two hours. The translator would come to my room every morning in order that we could go over the lecture to be given, and so that I could explain points not familiar to him. I had never given a lecture with the aid of a translator. There was one very good point about this that made giving the lectures easy for me. This was that, during the translation, I had time to think of what

Figure 5-9. The faculty and their assistants attending my class on coordination chemistry in 1979 at Nankai University in Tianjin, China.

to say next. The attendees (**Fig. 5-9**) were busy taking notes and seemed to be very interested in the lectures. As it turned out, their notes might not have been necessary because the lectures were taped and copies were made of my slides. Two months later, they sent me a paperback book of all my lectures with copies of all my slides. I could not read the Chinese, but the formulas, chemical equations, and plots were the same as I had presented in my lectures.

Every other afternoon, we held discussion sessions, for the Chinese wanted to keep their guests busy and get as much information from them as possible. I used the blackboard and did most of the talking during these sessions. I told them about our research, and about research being done by others in some western countries. Much to my surprise, there were a few questions about important research being reported in current journals. We talked about the articles they had read that interested them. After several such sessions where I did most of the talking, I said, "Now that I have told you about our research, let me hear from you about your research." Only three of the group came forward to describe their research. What one was doing sounded very interesting. She was following the electrolyte content in the blood of a person under acupuncture as compared to that of the person before having had the treatment. She was finding some reproducible differences with the largest being changes in the calcium content.

In addition to research, I also talked about teaching at both the undergraduate and graduate levels, and what courses our department required for a chemistry B.S. degree and for a Ph.D. degree. There was considerably more interest in the discussions on teaching than on research. Again, I asked them to tell me about their teaching in China—the courses required for a particular degree, as well as the course content. What they were doing did not seem too different from our teaching. However, I got the feeling that they were asking students to memorize many experimental facts about the elements and their reactions. This was being done at the expense of teaching them about the important theories of chemistry, such as bonding and reaction mechanisms. It was clear that when Chinese students arrived in the US, they were able to do the work explained in detail to them, but were less able to think their own way through a research problem.

As was true in Peking, we were invited to attend an event each evening. One evening, members of the Ministry of Education invited Mary and me to a banquet. This meant a large table with place settings for 10 or 15 of us, and a nice dinner, elegantly served. During the dinner, we had wine which was very sweet. Even though I did not like it, I managed to drink it. They also had a strong liquor, mao-tai, which had an odor of kerosene that I did not like, but drank. During the dinner, the Chinese would, at times, use mao-tai to make a toast, such as "for better relations between China and the U.S." I had been told that each time one of them gave a toast, I was expected to return it with a toast. My toasts were similar to theirs, such as "to a better understanding between the US and the PRC." After several of these toasts, we became more relaxed and friendly, making it easier to have an unguarded conversation on a variety of subjects. Because of the number of toasts, they decided I liked mao-tai and gave me a bottle of it to take home. Over the years, when we had guests for dinner and China was mentioned, I would offer them a drink of mao-tai. Most of our guests would either not finish their drink, or not accept a second glass when offered. The bottle was still almost full when we were invited to a Chinese wedding. Having been told that they would have a buffet meal, I decided to bring the bottle of mao-tai. Without anyone seeing, I placed the bottle at the end of the buffet table. The wedding was attended by young Chinese graduate students who wondered who brought the costly bottle of mao-tai. By the time the party was over, the bottle of mao-tai was empty, and I felt vindicated for not having to throw it out.

One other point of interest was when Mary and I, with a Chinese lady as interpreter, went in town to shop. Tianjin is not a tourist city, and, in 1979, there were very few tourists in China. Everything in China was very inexpensive by U.S. standards, and we bought many gifts for our children and grandchildren. As we stood around looking at items to bring home, the Chinese would come up to stare at us as if we were aliens from some other planet. Our interpreter said, "Just ignore this. You are the first Americans whom they have ever encountered and they want to see you and the clothing you are wearing."

One day we were told that we had a free evening. I asked if we could go visit the Chinese who had come to attend the lectures. They were in a near-by building, whereas foreigners occupied the main hotel. After dinner, Mary and I went over to visit with several of the people waiting for us. These were mostly those who knew some English, thus making communication relatively easy. We had taken with us a map of China, and pictures of our children and our home. We had them show on the map where they were from. Some came from as far north as Harbin, as far south as Guangzhou, and as far west as Lanzhou. Some had to travel by train, and because of the distance, slowness of the trains, and poor connections, it took a few of them two or three days to arrive in Tianjin. We asked questions about their families, and the nature of the cities in which they lived. A few gave us rather vivid answers, but several had little to say. They were all very excited about seeing the pictures we brought with us, and had many questions to ask. Mostly these questions had to do with life in the U.S. We left them feeling we were kind, interesting people wanting to have friendly relations with them. After our visit with them, it was clear that they felt much more comfortable around us during the remainder of our stay.

One morning, as I was walking toward the lecture room, I was stopped by a few of the older professors responsible for the lectures. They all were very excited about the good news they wanted to give me. They said that Vice-Premier Fang-Yi was told that my visit and lectures were a big success, and he wanted Mary and me to return to Peking to meet him. We were scheduled to go next to Shanghai, but the schedule was changed and we returned to Peking. We were to meet Fang-Yi in the Hall of the People (which corresponds to our congressional auditorium). We were told by our hosts that our visit with the Vice-Premier would take only about 10 to 15 minutes. Fang-Yi was there to meet us when we arrived. Two translators and three photographers were also present. Pictures were taken of us in front of a large portrait in the Hall of the People. Mary and I, the Vice-Premier, the translators, and photographers then went to a large room used for conversations with invited guests. Fang-Yi was a small, friendly, and intelligent young man. I immediately felt at ease talking with him, but later Mary told me that, as

Figure 5-10. Photograph of Vice-Premier Fang-Yi who, at the time, was the number three person of the Chinese government. He was in charge of agriculture and industrial development, along with other responsibilities. He invited me to come to their Hall of the People (corresponds to our Congressional building). He asked me many questions, but the one of most interest to him was "How long do you think it will take China to become an industrialized country as are other such countries?" I told him my guess would be about twenty years, and his response was "We cannot wait that long." It was November, 1979. Add twenty years to 1999, and my wild guess was quite close to accurate.

she sat listening off to the side, she became very nervous and wished the interview would soon end. However, it lasted a half hour instead of the 15 minutes we had been told to expect. He was president of the Chinese Academy of Science, responsible for the development of agriculture and industry. We sat in large chairs with our translators behind us, which reminded me of having seen Chairman Mao, President Richard Nixon, and Secretary of State Henry Kissinger in a similar situation on TV (**Fig. 5-10**). (It should be mentioned that President Nixon was extremely well liked by the Chinese because he was the American who restored diplomatic relations between the US and the PRC.)

At the time, as he must have known, I had recently been elected a member of our National Academy of Sciences (NAS). As President of their Academy, Fang-Yi started by asking a few questions about our NAS—questions about the number of members and the functions and responsibilities of our Academy. He would then volunteer the same information about their Academy. Next, came his questions of most interest to him: "Now, having visited several universities and their labs, what do you think about the level of teaching and research in China?" I responded by telling him that the teaching of chemistry was marginally satisfactory, but the faculty and members of their research groups were not able to do modern chemical research. I further told him this was not because their chemists were not capable of doing good research, but that they were not able. He asked what I meant by their scientists being capable but not able? I answered that they had the intelligence to do good fundamental research, but in chemistry they were not able to do modern research because they did not have the necessary instruments. All I saw at the better universities were facilities for gas chromatography, dropping mercury electrochemistry, X-ray powder patterns, and a 30 MHz NMR. I told him about the expensive instrumentation required to keep up with the basic research currently being done worldwide. The bottom line, I said, was that the government must make available the modern instrumentation and chemicals required if quality research was to be done.*

His next question could be seen to be even of much greater concern. He asked, "From what you have seen, how long do you think it will take China to reach the status of other industrialized nations?" I told him that I had no way of knowing, and that I was not qualified to give an answer. He kept pressing me on this until I finally said at least 20 years. His reply was, "We cannot wait that long, because now the prospect of China reaching its goal of becoming an industrialized nation looks very promising." As I think back, my guess of 20 years in 1979 seems close to the mark in 1999. We were told that Fang-Yi's visit with me was briefly shown on TV. The next day my picture was in their paper, with a short discussion of our meeting.

Five years later, Mary and I would remember our experiences during our month in China, where we were treated like VIPs and where we made many new friends. Shortly after we returned, I began to get letters from some of the people who attended my lectures. They all said kind things about my visit and lectures there, and asked if they could come to do research in my lab. I would have wanted to take any of them, but I had to respond that my lab was full and my resources were limited so that I was sorry but I could not be of any help. However, my presence in China for a month at a time when few, if any, chemists had yet been there resulted in their knowing about NU and they wrote to our department to ask about coming here as students.

---

*Five years later I returned to China after the World Bank had selected six of the major Chinese universities to receive funds for instruments and for chemicals. I visited three of these universities and found that the instruments were excellent, with the latest models of x-ray diffractometers, NMRs, infrared spectrometers, visible ultraviolet spectrometers, and mass spectrometers. They also had funds to purchase chemicals, as needed, for their research. Some of the universities had built a separate analytical services building. The instruments were mainly used for research, but some were used for teaching undergraduates. Unfortunately, the next time I was in China I found some faculty members frustrated because of the red tape required to get chemicals and to get instruments serviced. Yet, it was clear that the World Bank assistance had made a big difference in both their research and teaching.

Professors Qi-zhen Shi informed me that the PRC and the US had a program wherein the NSF's of the two countries would equally fund a collaborative project between PRC and US scientists. At the time (1983) they asked me about this, I was President of the ACS and said that we would wait until I had completed my Presidency to begin our collaborative efforts. They were on the faculty at Lanzhou University, so during the years of this collaboration, I went to Lanzhou a few times to prepare a paper on our research for publication. A few years later, our department and that at Lanzhou arranged an agreement that each year we would accept their very best student as one of our doctoral students. This arrangement has worked very well and we have had some excellent students as a result. In addition, Professor Liang-Nian Ji of Zhongshan University, who spent a year in my lab doing some excellent research, had me as a guest speaker at his university. He has successfully arranged some international conferences in China and invited me to come to give a plenary lecture but, for health reasons, I was not able to travel.

I have been to China seven times, and each visit had to do with our collaborative research, or with an invitation to a conference or a workshop. Each of these visits allowed a few days for sightseeing. As a result, I have seen all of the tourist sights, and other places not available to tourists. Each time when I was taken to a forbidden place, my host had to ask the authorities for permission. One of the more interesting places we went to was Hainan Island in the South China Sea near Vietnam. The island, which had only recently been made a new Province, has much the same climate as Hawaii, and a similar large beach. It seemed a shame that such a nice place should be forbidden and empty of even Chinese, except for some caretakers. Because of my many visits to China, one of the Professors at Fudan University said, "Basolo, you are the Marco Polo of chemistry to China."

When we returned to the city the next day, the Mayor arranged a banquet for us. He was a large Chinese, who did not know English and had a loud voice. He smiled a lot and had a good sense of humor, as he told us stories about the Island. He asked me if I liked Chinese food. I told him I did, and that I always preferred to eat the food of the country that I was in rather than our western food. When the soup was about to be served, he asked if I had ever eaten snake soup. The answer was no. The large bowl of soup was placed on the table before us. One could see pieces of snake floating in the soup. I forced myself to have seconds, and the Mayor could not believe it. However, I have had much better soup and cannot recommend snake soup. The Mayor said, "In Hainan we eat anything that moves."

My last visit to China was in 1997, when I was invited to a workshop-type meeting in Shanghai. This was arranged by one of our Chinese graduate students, George Yanwu-Li. He was able to get travel support from the National Science Foundations of both the US and the PRC. This was a unique meeting, intended to have only three main faculty lectures. The remaining lectures during the four day meeting were given by US and PRC graduate students. They were students who had almost finished their dissertation research for the Ph.D. It was interesting to see that the research studies being done in the PRC were on topics of current interest. The students of the two different countries had an opportunity to talk with each other about many topics of interest to them, such as life in each country. One of the subjects of their discussions was what jobs they would get after having obtained their degrees. This was during a period when the job market for Ph.D. chemists in the US was very low, so most of our graduate students were having to take two or more years of postdoctorate positions. Not so in the PRC, where there was a shortage of Ph.D.s in chemistry. However, when asked what positions they would have, they did not know because the government would appoint them to positions where they were needed.

This was my last visit to China and it was unbelievable what had happened in China during the 18 years since my first visit. Shanghai was building large high-rise buildings. There were now a large number of new hotels to accommodate the many tourists, large interstate-type highways were available, etc. This was largely due to the death of Chairman Mao, and the appointment of Deng Xiao-Ping as Premier. He was well known in the US because of his extended visit here, and because his daily activities were covered on TV. Most of us liked him, for he seemed to be enjoying himself, and to want to talk with the President and other persons of high status as well as ordinary factory workers and farmers. It appeared he was just the opposite of Chairman Mao, and some of us felt he would be making important changes in China—giving the people greater freedom. He started this immediately by allowing the farmers to sell a portion of their harvest on the open market. Some of the professors talked about them as being "rich farmers." Deng Xiao-Ping was responsible for many other important major changes. Unfortunately, as an elderly and ill man, he made the wrong decision on Tienemin Square, as is known worldwide.

I know many chemists in China, and we occasionally see each other at some scientific meeting. Professor Xiao-Zeng You of Nanjing University stopped to give us a seminar on his way to visit his son, a graduate student at the UI. You is the director of a large institute of coordination chemistry in Nanjing University, PRC. He is on the Editorial Board of the journal I edit, *Comments on Inorganic Chemistry*. China has been very generous to me. I received the Chinese Chemical Society Medal, and was awarded Honorary Degrees from Lanzhou and Zhongshan Universities.

GERMANY

I have been to Germany several times, with my host usually being Professor E. O. Fischer at the Technical University of Munich (TUM).[*] Fischer prefers that his friends call him E.O., and, as a good friend of his, this is what I will use. I visited Germany in response to invitations to give lectures on our research at various universities or at some international symposium. I cannot recall all of these visits, therefore I will limit my remarks to three of my longer stays in Germany.

---

[*]He and Professor Geoffrey Wilkinson shared the Nobel Prize in 1973 for their research on the cyclopentadienyl metal compounds.

One was in 1969, when I was a NATO Distinguished Professor at the Technical University of Munich, being hosted by E.O. I was there for a month, and gave lectures to students taking a course on organometallic chemistry. I managed to likewise give two talks on the courses I taught at NU, one to freshmen and one to graduate students. The students in Munich seemed to prefer my lectures on teaching, perhaps because I have always enjoyed teaching and this might have been apparent in my presentations.

Because of my interest in teaching, I wanted to attend a lecture in a beginning chemistry course given by Professor Egon Wiberg (**Fig. 5-11**) in the new chemistry building of the University of Munich. Professor Wiberg was noted for his teaching, as well as his research. He was able to design the lecture room to his liking, and get what he wanted for his large classes. (Note that a faculty member receives a bonus from the university proportionate to the number of students within his classes.) The room was very large, and the students came nowhere close to filling it. I sat in the back row to listen and watch his lecture. I do not understand German, but I was able to follow his lecture on silicon (Si) by observing the equations used for the syntheses and reactions of some silicon compounds. At times during the lecture and demonstrations, students would knock on their desks, making a loud noise. This knocking corresponds to our clapping. The lecture demonstrations were done by a staff member, hired particularly for this chore. The demonstrations were professionally done, timed perfectly to the lecture, and received the most knocking by the students. (All I could think of were our lecture demonstrations given by graduate students. Often the demonstrations would not work, and our students would get a kick out of the failure. However, when the reason for the failure was explained, the students seemed to learn more from the failed demonstration than from those that were successful.) The most spectacular thing about the lecture hall were the two large periodic tables, to the right and left of the front of the room. On the desk, there was a panel with controls that reminded me of a control panel on a B-747 airplane with Wiberg as the pilot who could push a button and light up the element that he wished to discuss. He could do the same for a period, for a row of elements, and was even able to designate diagonal similarities in the periodic table.

Figure 5-11. Professor Egon Wiberg (1901-76) of the University of Munich, Germany. His research dealt with hydrides of boron, aluminum, silicon, and gallium. He was known to be a great teacher of chemistry to beginning students, and to have designed a unique large periodic table in the lecture room.

On one occasion, I was in E.O.'s office when one of his students came to state that he had prepared a new compound. E.O. asked the student what he believed the compound to be, and then wrote something in his record pad and wrote a number on a label which he attached to the glass vial containing the compound. He gave the identifying number to the student. I asked him, "What is this all about?" E.O. told me that he would send the compound to an analytical lab to have it analyzed, and the results of the analysis would be returned to him. This record provided him a way to keep track of what his students were doing and to prevent them from making any mistakes. After our conversation, he suggested that we take a walk through his labs so I could meet some of the members of his research group. Most of them told me briefly about their work, and said how much they enjoyed my lectures. In one of the labs there was only one person, and, when he was introduced to me, he immediately stated in a rather loud voice, "I was waiting to meet you to tell you that your book on mechanisms is wrong." I responded, "I doubt that it is all wrong, but any mistakes were made by my coauthor Pearson." As you see, this is the advantage of having a coauthor. This student later came to the US and accepted a position as a research chemist but caused so much trouble that he had to move from job to job.

On one of my short visits to Germany, after E.O. had received the Nobel Prize, we were invited for dinner at the home of Professor Wolfgang Beck of the University of Munich. Wolfgang is a very good friend of mine and an outstanding inorganic chemist. That evening there was a discussion about plans to move TUM to Garching, a suburb of Munich. E.O. was strongly opposed to this, maintaining that it was too far from Munich and, without satisfactory public transportation, students would stay in the city and attend the University of Munich. He was adamant about this, asserting that he would not move out to Garching. Later, as a Nobel Prize laureate, he did stay in Munich and was provided an office in what remains of TUM.

Our next visit to Germany was in 1974 to attend the celebration of the 80th birthday of Professor Walter Hieber, the "father of metal carbonyl chemistry." The celebration was held in the very small and beautiful village of Ettal, located in the Bavarian mountains. The celebration was a big success with excellent talks and exquisite Bavarian food, beer, wine, and fine organometallic chemists. Through an interpreter, I was very pleased to meet Hieber and congratulate him on his 80th birthday and his life-long devotion to doing outstanding research on metal carbonyls. I told him that we had exchanged comments after his lecture in Amsterdam (pg. 45). He said that he recalled our exchange and that I had prompted him to read our publications on the subject. He said that he had found them of interest, and now agreed that some philosophy (theories of bonding and mechanisms) is needed in chemistry. One of my duties at this cele-

Figure 5-12. Professor E. O. Fischer and me having a friendly conversation while enjoying their good Bavarian beer.

Figure 5-13. Professor F. Albert Cotton of Texas A&M University at left, Andy Wojcicki, and Geoffrey Wilkinson (1921-96) in Ettal, Germany in 1974 to celebrate the 80th birthday of Walter Hieber.

Figure 5-14. Professors Fischer and Wilkinson demonstrating that even Nobel Prize Laureates are human, and can respond to friends enticing them on to dance the German jig.

bration was to introduce E.O. to give his lecture. During my introduction, I pointed out that although Fischer and Wilkinson carried on a bitter polemic in the literature regarding their research, they both had agreed to let by-gones be by-gones and to jointly accept the Nobel Prize in 1973.

Ettal was so small that there was nothing to do in the evening. However, our hosts had prepared for this by having a bountiful supply of good Bavarian beer (**Figs.** 5-12,13) and a German band. After consuming enough beer, some of the people started to dance. At one point, there was a lot of laughter and clapping to try to get E.O. and Geoffrey to dance German-style together. This they did, and a very nice picture was taken of the two Nobel Prize Laureats dancing the jig together (**Fig.** 5-14).

One evening, Professor Lamberto Malatesta of the University of Milan and his wife were trying to interest some couples in playing a game of bridge. Finally, my wife Mary accommodated them by agreeing to play. Mary and I never played serious bridge, but the Malatestas *only* played serious bridge and had accumulated some master points. Therefore, the outcome of the game was obvious, yet we declared that this would be a match between Italy and the US. Some of the other people, not having anything better to do, gathered around to watch us play. As luck would have it, Mary and I continued to get good cards. The hands dealt to us were such that we won and upheld the honor of the US.

In 1993, I was awarded a Senior Humboldt Fellowship with Professor Edward W. Schlag of the TUM as my sponsor. The Award was for six months, but Mary and I could only stay for three months and were given the option of returning later for the second three month period. Neither Mary nor I know German, but none was really necessary because most of our friends were from the universities and spoke English, as did the persons at the hotel desk. We lived in a three room dwelling, with a kitchen, a living room, and a bedroom, only a 15 minute commuter train ride to the center of Munich. We were near large department stores, restaurants, and a self-serving supermarket which made it easy to shop for groceries. Ed Schlag and his wife Angela made certain that all our needs were cared for and that we felt comfortable with our living quarters. We had all known each other when Ed was a Professor in our department. Born in Germany, he could not pass up the opportunity to return there when offered a Professorship at TUM. At about that same time, two other of our faculty with German backgrounds also accepted Professorships at TUM.

The Senior Humboldt Fellowship allows its awardees to do whatever they wish in research or in writing. It also encourages awardees to interact with faculty at the host institution, and to spend some time visiting other universities in Germany and Europe to give lectures and to establish an awareness of the programs of the Alexander von Humboldt Foundation. During the three month stay, I gave lectures and visited inorganic chemists in ten different Germany universities. Likewise, I was invited to visit and give seminars in Denmark, France, Italy, and Poland.

Since Ed is a physical chemist and I am an inorganic chemist, we differ markedly in our areas of research. This meant that at the TUM I spent my time mostly reading research progress reports sent to me by members of my research group in the US, which also meant that I was preparing manuscripts for publication at the same time. In addition, I was fortunate that the TUM has two of the very best inorganic chemists in the world, Professors Hubert Schmidbaur and Wolfgang A. Herrmann. Their students and assistant faculty would make appointments and come individually to tell me about their research. The work they were doing was extremely exciting to me and near enough to our research interests that we were able to have some meaningful discussions. Furthermore, I was asked to give a seminar on our research, after which some of the persons I had seen earlier wanted to return to talk with me about certain points that I had made in my lecture. In addition to the research being done at the TUM, I likewise was able to hear about research being done by the research groups of Professors Wolfgang Beck and Hans Nöth, both outstanding inorganic chemists. Wolfgang's chemistry on the syntheses of metal complexes was close to our interests, whereas the research that Hans was doing on boron chemistry was not. He and I, however, did have something in common that we could talk about—he was President of the German Chemical Society and I had been President of the ACS. We compared notes to see how the two societies were similar and how they differed.

Shortly after we arrived in Munich, the Humboldt Foundation was having its two day annual meeting, and the Humboldt Awardees were invited to attend. At one of the dinners, Mary and I were sitting at a table with Professor Wolfgang Krätschmer of the University of Heidelberg and his wife. Even though he is an astrophysicist, I knew of his research because Professor Richard E. Smalley (Nobel Prize, 1996) of

Rice University in Houston, Texas, had discovered the bucky ball C_{60}. This observation, in and of itself, though important, would not yield any concentration of the material with which to do chemistry. However, Wolfgang noticed that the spectrum obtained by Rich was similar to what he had observed in outer space. Wolfgang then began to try to make C_{60}, and he finally succeeded. His synthesis gave quantities of C_{60}, making it possible to do chemistry on it, and rapidly generating thousands of scientific publications worldwide. I later told Mary that we were at a table with a person destined to win the Nobel Prize in chemistry. I was proven wrong in 1996 when the Prize did not include Wolfgang, but in my heart I think that I was correct and the Nobel Prize Committee was wrong.

Mary, who had quadruple heart surgery a few years earlier, was having some problem with her heart, which was taken care of by an angioplasty. Still, our children wanted her to come home, so she returned about one month before I returned. Thanks to Ed and Angela, Mary and I had a wonderful time in Munich. I personally thank Ed for his effort in making it possible for me to be awarded the Humboldt Fellowship.

AUSTRALIA

I have been to Australia four times: once to attend an IUPAC (International Union of Pure and Applied Chemistry) meeting; once as a visiting lecturer for one month at Monash University in Melbourne; once as representative of the ACS when I was its president; and once when I was invited to introduce the first speaker at an ICCC meeting.

The IUPAC meeting had many scientific presentations that interested me, and I was young enough to feel that I had to attend all of the talks. We were there long before the now landmark opera house was built. One day, my colleague Professor James Ibers and I rented a car and drove out in the country, wanting to see some kangaroos and koalas in the wild, and perhaps some aborigines. It started to rain, so we spent most of our time in one of the outback pubs enjoying the Australian Foster beer.

In 1976, I was invited to be a Visiting Professor at Monash University in Clayton, Victoria, near Melbourne, to give a series of lectures in a course on inorganic coordination chemistry. The faculty member teaching the course had been appointed to an administrative position, and funds became available to have four visiting lecturers —one each month for the four month course. The other lecturers were Professors G. Wilkinson, Otsuka, and A. M. Sargeson. My one month period was from mid-November to mid-December. This is their summer, but yet they were preparing for Christmas. Coming from Chicago where snow—at times, much ice and snow—is the norm around Christmas, it was a strange feeling to go to Melbourne City Center and see their large Christmas tree with all of its lights.

Although I was never more than a "duffer" golfer, I did enjoy playing. Some of the chemistry faculty at Monash University likewise enjoyed golf, and we managed to play a couple of times a week. There are a large number of fine courses near Melbourne. I was told that the Melbourne area, centuries ago, was engulfed by the Tasmanian Sea and that underneath the top soil, it is covered with sand. This means that sand traps are readily built on their golf courses, and most courses have more than the usual number of such traps. I will always remember a par three hole I played had a very large and very deep sand trap close to the front of the green. The fellows with whom I was playing cautioned me not to get in the trap, because I would never be able to hit the ball out. It was a rather short par three shot, so I tried to hit a high ball on the green. My shot was straight at the green, but, unfortunately, it was short and fell in the deep trap. The fellows with me said, "You will never hit the ball out, so just throw it out and take a penalty stroke." I did not take their advice and decided to hit the ball as hard as I could with my sand wedge. We were all surprised (mostly me) when the ball made it over the edge and rolled down near the hole for a gimme par three. Golfers will understand why this was a high point of my month at Monash University.

In 1976, I was awarded the Frank Dwyer Medal, an Award that I am very pleased to have received, because Frank and I were the best of friends (see pp. 183-186). In 1983, I was delighted to be asked by the ACS to make a tour to New Zealand and Australia to give lectures at the Local Sections of their chemical societies. Their sections meet mostly at universities, which gave me the opportunity to visit their labs and talk with them about their research. My first stop was Auckland, which is often a stop made on flights from the US to Australia. There my host was Professor Warren Roper, an outstanding organometallic chemist, at the University of Auckland. I was amazed when he told me that he only had five students in his group doing research, and two were undergraduates. Because he is one of the most important organometallic chemists worldwide, I asked him how he could compete with other outstanding chemists who had research personnel of 20-30 graduate students and postdoctorates. He replied that it was because he worked on new problems, made new compounds, and discovered new reactions. Once his new work was published, it pleased him to see others continue the research, helping him to become famous. He said that he would then move on to an entirely different problem.

The other stops on my schedule were reached by travel in a small airplane. The airplane flew at a low altitude, making it possible to have a good view of the green country below. Everywhere there were white sheep grazing, which outnumber the population of New Zealand. My hosts at the University of Victoria in Wellington were Professors N. F. Curtis and D. C. Weatherburn. In addition to my usual discussions with some of the faculty about their research, I was asked to have a talk with their Secretary of Commerce. He had authorized the building of a large plant to convert methane (CH_4) into methyl alcohol (CH_3OH) and then into gasoline. Because New Zealand has a small population, the cost of the plant would require a large percentage of the country's budget. Furthermore, no such plant had yet to be built elsewhere in the world, so many of the people were opposed to the plant for fear it would not work. The Secretary wanted to ask me, as a chemist, if I believed it would work. I told him that I was an inorganic chemist and knew little about petroleum chemistry. He kept pressuring me to say if I thought the process had a good chance of being successful. I finally told him that I had read that Mobil Oil had discovered a zeolite, ZsM4, which catalyzed the conversion of methane into methyl alcohol. I further told him that there were well-known processes to convert methyl alcohol into gasoline. The next day I was quoted as a famous US chemist who said that the plant would be a big success. I am told that the plant is working and producing gasoline from methane, which they get in abundance from a natural source.

I then went to Christchurch where my host was Professor D. A. House at the University of Canterbury. Don and I had known each other for years because our research had to do with the kinetics and mechanisms of ligand substitution of Werner metal complexes. As he had continued to do work in this area, we had much to talk about. Also on the faculty at this university was Professor W. T. Robinson, who had spent some time at NU working in Jim Ibers' lab. One of his graduate students showed me their X-ray lab.

My tour ended in Dunedin at the University of Otago where my host was Professor D. A. Buckingham. We knew each other for several years, and he visited me at NU where he gave a seminar on his research. One thing that interested me was the weather—as we went south, the temperature became colder, unlike the US, where the temperature gets warmer as one goes south. In fact, Dunedin is as close to the Antarctica as I would ever get. There was snow and ice in Dunedin, whereas northern New Zealand had good, comfortable sunny weather.

I then went to Australia, which I had visited twice before, so I will just briefly mention where I went to visit and give lectures, and say something about a few of them. I first went to Hobart in Tasmania where I was shown the ferocious Tasmanian devil (dog). Then to Melbourne, and to Canberra where my hosts were Professors J. A. Broomhead and A. M. Sargeson at the Australian National University. John had been a postdoctorate in our lab for a year, and I knew Sargeson from his published research. The University arrangement was most unusual, in that part of it was a research center with excellent funding, which produced some of the very best research done in Australia. Alan was on the faculty in this part of the university. Teaching was done by another branch of the school where the faculty had less time for research. John was a faculty member of this part of the university.*

---

*It is interesting that, when at NU, Broomhead studied the chloride ligand replacement reaction of $[Cr(NH_3)_5Cl]^{2+}$. He made a detailed kinetic study of its rate of hydrolysis, and proposed a possible mechanism. Professor Allen of the University of Toronto, Canada, told me that his student, Senoff, had independently made essentially the same study and obtained the same results. This was to be part of Senoff's Ph.D. dissertation, when he found our publication in the current literature. Unfortunately, this meant he would have to work on some other problem to complete an acceptable dissertation. Since the synthesis of the Cr complex involves a complicated reaction that was not understood in detail, they decided to investigate this reaction. The long and the short of it is that, as a result of this investigation, they made the discovery that N_2 can behave as a ligand—in this case, to give $[Cr(NH_3)_5N_2]^{3+}$. This discovery stimulated research worldwide and the publications of thousands of scientific papers. One of the goals in these studies was for chemists to find a catalytic way to cheaply and efficiently convert coordinated N_2 into NH_3. Nature does this by using the enzyme nitrogenase, and chemists, now knowing that N_2 could function as a ligand, hoped to be able to better

In Sydney, I was able to spend time with Professor James K. Beattie, one of my former graduate students, who holds a record for having obtained his Ph.D. at NU after 2½ years. He was an undergraduate student at Princeton University, where he also swam on its varsity swim team. He then became a Rhodes Scholar, received his Masters Degree at Cambridge, and then obtained his Ph.D. at NU in 1967. He went directly to a faculty position at the UI. There he met Margereth, who was working on her Ph.D. in history. They were married and went to Sydney, where they both joined the faculty of the University of Sydney. Jim and I enjoyed reminiscing about his graduate days, and I met his wife and their twin children (one a girl, the other a boy). Jim filled me in on his research which was going very well, and which came as no surprise to me.

Next I went north to the University of Queensland in St. Lucia, Brisbane, where Professor J. D. Cotton was my host. After my seminar, he and I talked about metal carbonyl chemistry that we were both engaged in. He showed me much of Brisbane, after which I went to Townsville near the Barrier Reef where my host was Professor L. F. Lindoy at North Queensland, James Cook University—now at the University of Sydney. He invited me to stay as a guest in his house, where I met his wife and children. This was a pleasant stay, because I enjoyed talking with his family and because it gave me a break from always having to stay in a hotel. His work on determining the stability constants of many different metal ligand combinations was of interest to me, so he told me all about his research. Since he was not doing organometallic chemistry, he was less interested in our work. We did not go to the Barrier Reef, but we did go to an aquarium that had many of the same fish and corals that were at the reef. The same environment was provided by passing sea water through the fish and coral tanks.

Next was my first and only visit to Western Australia. My host at Perth was to be Professor B. N. Figgis, but at the time he was at the Argonne National Laboratory near Chicago. Instead, his wife had several of his colleagues and spouses join us for dinner. It was a fine dinner and we all had a delightful time. I knew that the Australians liked their Foster beer, but I did not know how much they could consume in one evening. The day following my talk I was taken sightseeing in and near Perth. It was of interest to me to have completed my desire to go across the large country of Australia.

Each of the first three times it was not possible for Mary to accompany me, because of family problems with children at home. Mary now had many friends from Australia who told her about it, and she had also read several books about Australia. The third invitation I had to visit Australia was in 1989 to attend the ICCC in Brisbane. At this stage, our four children had married, leaving Mary and me home alone. Mary could now come with me, but the question was whether I would be able to travel to Australia. I had severe back pain, and it seemed that surgery might be necessary. I asked our doctor

---

understand how the enzyme works. Neither of these goals has yet been reached, but the effort made has produced some useful new chemistry. It is sad that Allen died at such a young age and could not receive the full satisfaction of his most important discovery. Broomhead, Pearson, and I had the satisfaction of knowing that we indirectly had something to do with the discovery that N_2 can function as a ligand in certain metal complexes.

and personal friend, Bernard Adelson, if it would be OK for Mary and me to go to Brisbane. I told him how much Mary would like to go, and that this might well be our last chance. He said, "OK, if you can stand the pain, and we can take care of you when you return." When we arrived in Australia, my pain got worse, requiring that I stay in bed much of the time. Fortunately, we had personal friends who took Mary sightseeing while I remained in the hotel room taking care of myself. On our way home, we stopped in Hawaii where I tripped on an escalator and fell. I was not badly hurt, but when I got home I immediately had surgery on my back. That was the first of five back surgeries. In spite of the pain during my last trip to Australia, I was glad that I had made the effort to go because Mary had the chance to get to Australia—the country she had wanted to visit for years.

KUWAIT

In 1984, I had a kind invitation from my friend Professor M. S. Al-Obadie and Dean of the Faculty of Science Ali A. Al-Shamlan to be a Visiting Professor for two weeks during early December at the University of Kuwait in Kuwait City. I wanted to accept their invitation because I had never been to that part of the world, nor had I ever been in their type of culture. I told my family of the invitation and we discussed whether I should accept it. We were troubled because a few months earlier there had been a disastrous highjacking of a Kuwait airplane. My family felt it was not safe, and I should not go. My feeling was that this would be the best time to go because, after the highjacking, security would be at its highest. Since my family could see that I wanted to go, they reluctantly agreed that I could go.

During my two weeks in Kuwait I was to teach a coordination chemistry class to 18 students. Many things of interest to me happened during the two weeks. The first of many experiences was receiving a first-class ticket on a Boeing 747. I had never flown first-class for such a long distance, and I must admit that I enjoyed the elegance of the trip. (Later I learned that my entire stay in Kuwait was all first class.) The flight was nonstop from New York to London. There the pilots and stewards were replaced by a new crew, and the flight continued nonstop from London to Kuwait. Upon landing in Kuwait, from the plane window we could see that there were emergency vehicles along both sides of the airplane—a fire truck, an ambulance, and a military jeep. This seemed like an unusual welcome, but we assumed it was just high security resulting from the recent highjacking. Inside the airport, there were several military personnel, heavily armed, walking around. My host was waiting for me and, as I had not checked any luggage (the way I always travel), getting through customs caused no delay. I was taken to the Sheraton Hotel to get a good night's sleep.

My first morning at breakfast I introduced myself to some of my fellow passengers and we had breakfast together. They said that they had all been delayed for about six hours before they could claim their checked luggage and leave the airport. The reason for the delay was because the flight captain had received a message that there was a bomb on the airplane. This required a complete check of the plane, including the luggage.

I learned more about this bomb alert two weeks later upon my return flight. I was seated near a flight captain of the Kuwait airlines. He was on leave and on his way home to London. I told him about the bomb scare, and asked what had happened. He told me there was no bomb, and explained what had occurred. At the stop in London where the crew had changed, a few people entered to clean the airplane. One of the stewardesses who had gotten off the plane later discovered that she had forgotten her purse. She returned to get it and noticed one of the cleaners was cleaning the overhead closed luggage containers. When she got to her hotel, she began to think that what she had seen was not something normally performed by the cleaning crew. She telephoned the pilot and told him what she had observed. He felt that this was sufficiently serious to let the airport security know about it. The thorough check of the airplane found no bomb, but did find the problem responsible for the scare. Up behind the airplane sealing, they found some Playboy magazines. The pilot said that this and other types of contraband are always smuggled into Kuwait, since such "porno" literature is not allowed in the country.

Since I had never before been to a country like Kuwait, almost everything I encountered was of interest to me. For example, my first day at the University I was surprised to find that there were about the same number of female and male students (**Fig. 5-15**). I was of the opinion that women stayed mostly at home, wore black clothes, and covered their faces. Instead, the female students dressed normally, and even drove their own cars. I was told that this *was* unusual, for most other such countries maintain the old traditions. The eighteen students in my class were very young women and men. I enjoyed teaching them, and they kindly told me how much they enjoyed my lectures and our discussions.

Figure 5-15. Female and male Kuwaiti students.

The University administrators were all Kuwaiti, but some of the faculty were from other countries. I got to know three chemistry faculty members who were not Kuwaiti: one from Iran, one from Sudan, and one from Russia. This same situation existed in all other types of work. At the banks, the tellers are not Kuwaiti, while all the higher positions are held by Kuwaiti. The same is true for all executive positions, whereas the manual labor is all done by foreigners, mostly Indians.

When I go to a country that I have not visited before, I always like to buy a little gift, made in that country, for my children, grandchildren, and various staff members at NU. Therefore, I went to a large elegant department store looking for anything made in Kuwait. What I found was that all such items were made elsewhere. I was told that Kuwait imports all its goods and food, exporting enough oil to more than pay for this. I was taken to an oil port where the huge oil tankers are filled for export. The Director of this operation (Kuwait National Petroleum Co.) showed me around and explained what things were and how they functioned. He then invited my host and me to have lunch with him and a few of his colleagues. We had an excellent lunch, without dry martinis, for alcohol is forbidden in Kuwait. During lunch we had casual friendly conversation, and, at some point, I mentioned that I liked to play golf which would not be possible in Kuwait. One guy said, "Oh, yes it is, and I play once a week." I asked, "How do you keep the greens green?" He replied, "We have smoothly rolled sand 'greens.'" Later he showed me the course with sand tee offs, fairways, and greens. I told him that I would like to play it, but only once. After lunch we moved to another room to have tea and continue our conversation. I recall saying to the Director, "Should not Kuwait spend some of its oil money to build factories for when the oil is all gone?" His reply was that it would last longer than my lifetime, so this should not be a concern to me.

During my two weeks there, I was invited once to dinner at the house of my host Al-Obadie. He had also invited several of the university administrators. We were in a large room with cushion chairs side-by-side, backed up against the walls of the room. We helped ourselves to the food on a table near the kitchen. When something was needed, it was provided by his son, and we never got to see his wife. Since we were seated in rows side-by-side on opposite walls of the room, it meant you almost had to shout to be heard when trying to converse. During the evening, I was presented with a gift of a Kuwaiti robe and a red and white plaited head scarf (**Fig. 5-16**). I immediately put the gifts on to show them how I would look as a Kuwaiti. I also put it on when I got home, confounding the children and grandchildren.

During my stay, two of the foreign faculty members also invited me to their house for dinner. I had a good Russian meal prepared by the wife, and a few drinks of black market vodka provided by the husband. I told him that I always had a glass of wine with my dinner at home, but when traveling and eating at restaurants, I would start with a dry martini and then have wine with the meal. My friends said that since there was no alcohol in Kuwait, I would have two weeks to test my dependence on alcohol. If I were an alcoholic, I would get the shakes which, fortunately, I did not. The other faculty member was Professor Yousif Sulfab of Sudan who came in 1987 to spend his sabbatical year with me as his host. I was pleased to meet his very nice family, and to spend a very delightful evening with them. They came to the US to stay when the Iraqis invaded Kuwait, fortunately getting away with their lives, but losing their house and all of their belongings.

Iraq's invasion of Kuwait reminds me of my encounter with their military. When I arrived in Kuwait, they had a car and driver at my disposal. He would pick me up each morning to take me to the university. One day, he and my host asked me what I wanted to see. I said that since I had never seen such a desert, I would like to see it up close. The driver said that he knew just where to take me. He knew a nomad who lived in a tent with his group, which was not far from the city. We went there and were invited to sit on some rugs on the sand and have tea. The women prepared the tea and gave it to us, but then returned to their side of the tent. I could not understand what the driver and the nomad were saying, but at times they laughed to indicate they were enjoying their conversation. My driver volunteered the information that the nomads in the group had multiple wives, and chose one each night with whom he would sleep. When we left, I asked the driver to get off the main highway and take us further into the desert. He seemed apprehensive, but proceeded to drive us over the desert sand while telling me about the desert weather. Once we got out some distance, he said we had better return. It was soon apparent why he was so concerned about returning to the highway to Kuwait City. We had crossed the border into Iraq, and the Iraqi military posted there stopped us, and took us to their quarters to interrogate us at length. None of them spoke English so I had to answer their questions through the driver. They looked at my passport, and wanted to know what an American was doing illegally entering their country. I told them that I was a visiting professor at the University of Kuwait, where I had been teaching a course for one week, and was to stay and teach it for another week. After considerable discussion among themselves, they made a few telephone calls. Finally, they believed me and said, "Get on the highway back to Kuwait, and never again enter Iraq."

Figure 5-16. This is me in a gift of a Kuwaiti robe and head scarf. It was given to me during dinner at the house of my host, Professor M. S. Al-Obadie; then he and his guests asked me to put it on so they could see how I would look as a Kuwaiti.

Needless to say, I watched the invasion of Kuwait by Iraq on TV—how they ruthlessly killed civilians and destroyed the beautiful city buildings and the museum, looting the buildings of valuable items. I was particularly sad to know that they had taken all the instruments and chemicals from the chemistry labs. The aggressive response of President George Bush, with the support of several countries, meant a great deal to me. He mounted the Gulf War, which immediately got the Iraqis out of Kuwait, winning the war with a minimum loss of US soldiers. It is puzzling why people continue to destroy one another, for whatever the reason.

My two weeks in Kuwait were a rewarding experience. On my return, even my family agreed that it was good that I had taken the trip. In 1989, I was nominated as an External Assessor of the *Journal of the University of Kuwait* (Science). This meant I was to return to Kuwait for a stay of three days to be a member of a panel to evaluate their journal. A good friend of mine, Professor Afif M. Seyam at the University of Jordan, asked me to please also stop to visit their chemistry department and give a seminar. I was there for three days and had a delightful time seeing the beautiful city of Amman, and also several of the tourist sights. Even of more interest to me was meeting the families of Afif and of Professor Homdallah Hadali and being a guest for dinner in their very nice stone homes. Afif was the first student of my colleague Professor Tobin Marks, and Homdallah was a student of Professor Duward Shriver.*

It has been a worthwhile pleasure for me to visit Israel, Kuwait, and Jordan. Now I have a better appreciation of some of the problems confronting the area.

* * *

There are other countries (see Appendix) where I spent less time, but have had some interesting experiences and made some very dear friends. I thank my hosts and others who helped make me feel at home in their respective countries.

---

*Afif had a Fellowship from Jordan to come to do his research with me. I would have liked to have him join my research group; however, I had just accepted a three year appointment as chairman of our department and my research group was large. Before accepting the chairmanship, I told the Dean that he must approve an inorganic faculty position. This he did, so I contacted Professor F. A. Cotton at MIT to ask if he might have an outstanding student who was looking for an academic position as an inorganic chemist. He said that he had just such a student who was delaying writing his dissertation to avoid getting drafted in the Vietnam War because students who had not completed their degrees were not subject to the draft. I said, "If we find that he is as good as you say, we can wait a few months." This student was Tobin Marks. We made him an offer and he indicated that he would be at NU in about two months. Therefore, I told Afif that he should wait until this bright new faculty member arrived and choose him as his mentor. I further told him that he could spend his time in the library reading Tobin's published scientific work, and that when Tobin arrived, Afif could start by helping him set up his lab. All of our plans worked out even better than expected. Afif and Tobin did the seminal work on f-element organometallic chemistry. The rest is history, with Tobin now being one of the very best inorganic chemists in the world. Afif continues to collaborate with Tobin on research, and he often comes to work in Tobin's lab for a few months or even a year.

6

FOREIGN GUESTS HOSTED

I start with the apology that my memory is not what it should be, so I may not recall all of the guests that visited us during my fifty plus years on the NU chemistry faculty. Furthermore, I cannot remember some of the dates of their visits, so few dates will be given. The order of the visitors will be alphabetical, by country.

J. K. Beattie

Professor James Beattie, of the University of Sydney, obtained his Ph.D. at NU with me as his mentor. Years ago, Jim visited us and gave a seminar on his research in progress, dealing with electron transfer and spin changes in certain metal complexes. He was in the US and the UK on a program whereby Australian faculty are supported for three months abroad, because they consider themselves isolated and in need of outside contacts. See the section on Australia for more information on Jim.

F. P. Dwyer

My visits to Australia were all of interest to me, but what I think that the members of our chemistry department enjoyed most about Australia was when the outstanding Australian coordination chemist, Professor Frank Dwyer (**Fig. 6-1**) of the University of Sydney, spent 1953-54 in our department with me as his host. All of this began when my group became interested in investigating the rates and mechanisms of racemization of $[Fe(bipy)_3]^{2+}$ and $[Ni(bipy)_3]^{2+}$. This research was done in collaboration with my former colleague Professor Henry Neuman, now at the Georgia Institute of Technology, and our student, Lester Seiden. Before my student was able to work on the problem, he had to first make the complexes and then resolve them (separate their d-ℓ, dextro-levo, right hand-left hand isomers).

The synthesis was easy, but the student was not able to separate the isomers using the method reported in the literature. After several attempts had failed, I said, "Although I do not know the author of the article, I will write to him and see if he can help us." The author of the article was Frank Dwyer. His quick response was most gratifying, for he sent us detailed information on how to resolve the compounds, along with samples of the optically active compounds. In his letter, he went on to say that he had never been out of Australia and that he would like to come and work in my lab. He also wrote that the importance of his research was not appreciated by his department head.

What funds I had for research were all committed at that time, but the department arranged to pay Dwyer to teach my graduate course on coordination chemistry. I wrote to Dwyer, telling him what our department could do and that I knew it was not much money but, nevertheless, it would be enough to get him through the year. Dwyer replied that he would accept the offer and come alone, feeling sure that he could manage on what the University was to pay him.

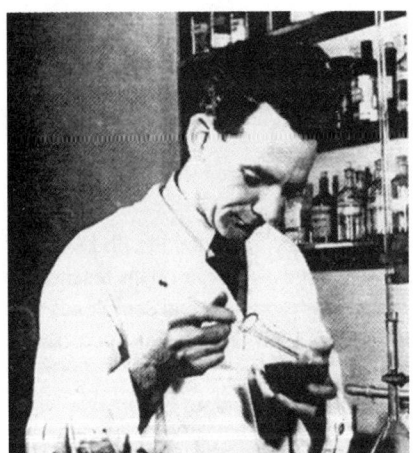

Figure 6-1. Professor Frankie Dwyer (1910-62), of the University of Sydney and, later, at the Australian National University in Canberra, was a hands-on inorganic chemist until the day of his untimely death. He was a superb coordination chemist, noted for his work on metal complexes with sexadentate ligands, his ability to separate optical enatiomers of chiral complexes, and his research on bioinorganic problems before the name had been introduced.

We understood quite a bit about his finances when we discovered that he had traveled on a freighter which was taking lobster tails to Boston. Why lobsters needed to be freighted to the northeastern coast is something we have yet to understand. Dwyer said there were ten passengers on the ship, but the captain was more concerned about his cargo of lobster tails than about his passengers. Frankie, as he was called by his friends in Australia, and very soon by his friends here, then took a train from Boston to Chicago. I was to meet him at the station, but since we had never met before, I was somewhat apprehensive as to whether I would be able to identify him. As I waited at the gate, I saw a small young man, dressed somewhat differently than the other passengers, loaded down with luggage. Sure enough, I had located Frankie with no difficulty at all.

Frankie was a guest at our house for three days while he looked for a room near the campus. He found the ideal place in the house of a widow whose husband had been a well-known physicist on the faculty at NU.

Frankie was given the run of the house, and the thing he liked best was the refrigerator where he could keep some beer—Australians like their Foster beer.

One morning during his three day stay with us, Mary and I were in the kitchen waiting for him to come down to breakfast. We looked at one another, both of us thinking the same thing: "Do you think we will ever understand him?" It took some time, but we learned that his cockney accent pronounced the letter "a" as if it were the letter "i", thus he would get up and "shive" in the morning and later pick up his "mile." Once Mary and I figured this out, we were OK and had no serious problem understanding him.

However, once Frankie started his stay in our department, the same problem arose with my faculty colleagues and others who talked with him. It therefore came as no surprise that, after a week of lectures in the graduate student course on coordination chemistry, a delegation of two or three students came to talk to me about the course. They said that Dwyer lectured much too fast and, even worse, they were unable to understand him. I explained to them that they had to try to change "a"s to "i"s in his words, and I also told them that I would talk with him. I did as promised, and suggested that he should try speaking slower and practice pronouncing "a"s as "a"s, not only for the students' sake but also because he had received invitations to be a guest speaker at other universities. His response was, "Basolo, you old bastard, you're just putting me on."

A week or two later he came in my office and said, "Fred, I owe you an apology." I did not know what this was about, but by that time he had started to do hands-on research. His lab was on the third floor, and the chemical storeroom was in the basement. He said that he would go down to get a chemical, such as sodium nitrate ($NaNO_3$), but when he was in his lab about to use it, he found that he had been given sodium nitrite ($NaNO_2$). He said, "This is a damn nuisance that I cannot get the compounds that I want" and then recalled that I had warned him about this. I went to the blackboard in my office and wrote

$$Na_2SO_4 \qquad Na_2SO_3$$

and asked him to tell me the name of each. The first is sodium sulfate; the second is sodium sulfite. As Frankie said the words, they both sounded the same to me—sodium sulfite. He insisted that they sounded different and told me what to listen for. I listened, but to this day I cannot detect any difference.

Dwyer was such a "funny," charismatic person that one could not help but like and enjoy him. For example, at that time he was best known for having prepared interesting and useful sexadentate ligands—ligands that attach to metals at six positions (**Fig. 6-2**). One evening, after a few beers, I teased him about the nomenclature of his new ligands. I said he used sexadentate instead of hexadentate to attract the attention of the readers. He said, "Fred, you stupid bastard, one should not use hexa which is Greek with the Latin dentate. The sexadentate word has Latin with Latin and is correct."

One other amusing happening was the day when Frankie came to our conference room where our chemistry faculty have their brown bag lunch each day. He announced that he had received an invitation to give a lecture. We asked, "Where?" and he replied "At Wine Stite." We had to ask, "Where in the hell is that?" We finally figured out that he was to go to Wayne State University in Detroit. This Dwyer saga could go on for several more pages, but it is now necessary to move on.

During Frankie's stay with us, one of his faculty colleagues, Professor Lions, stopped in on his travels from Australia. He spent three months working with Frankie on their joint research. Lions was an organic chemist who prepared the ligands they used in their research. Lions was easily understood, but he liked to joke about how our English differed. One thing he said is that before coming to the US, he believed a "basin" was a four legged animal, only to find that here it is something in which one washes their hands and face. One can see from my saga about Dwyer that I cherished receiving the Dwyer Medal Award in 1976.

In 1962, Dwyer was on his way to the Stockholm ICCC, and he stopped at NU to visit some of my colleagues—I was still in Rome on sabbatical. He received an urgent message that his son was near death as the result of an accident, so Dwyer immediately returned home, but while there, he died an untimely death. He was a chain smoker which may have contributed to his death. He and I had been in contact, and we planned to get together at the ICCC. When I arrived in Stockholm, Ron Nyholm was there to meet me, wanting to be the first to give me the sad news of Frankie's death.

Figure 6-2. Character sketch of Frankie Dwyer inside one of his sexadentate ligands (Courtesy of his student Alan Sargeson).

F. Lions

As mentioned above, Professor Francis Lions (**Fig. 6-3**) did his dissertation research in organic chemistry. When on the faculty at the University of Sydney, he became good friends with his colleague Dwyer. This was understandable, since they both had a good sense of humor and a very similar general approach to life. Lions became so interested in the research being done by Dwyer that he began preparing multidentate organic ligands that Dwyer could use to prepare complicated metal chelate compounds, where the organic ligand is attached to six coordination positions on the metal. Lions is best known for his synthesis of chelating agents important to coordination chemists.

Lions and I discussed our research, but we likewise talked about athletics. He was a successful sportsman, excelling in swimming, diving, cycling, and football. I played baseball, golf, and tennis, but certainly did not excel in any. I was on NU's faculty athletic committee, and was also its chairman. Frankie and I had much in common other than chemistry, and the two of us became the best of friends.

A. M. Sargeson

Alan Sargeson, Professor at the Australian National University in Canberra, obtained his Ph.D. with Frank Dwyer. Like his mentor, Alan is known globally as one of the very best coordination chemists. He was elected a foreign member of the US NAS in 1994. He, his wife, and small child visited us many years ago and I do not recall the topic of his seminar, but I feel certain that it was on the syntheses and reactions of some unique metal complexes. More recently, he attracted the attention of coordination chemists with the syntheses of what he calls encapsulate (**28**) because the ligand is designed to completely encapsulate the metal. This gives the metal complex a unique structure, with the metal trapped inside the coordinated ligand—properties not found with classical metal complexes. Alan has had the good fortune of being a professor in the research part of the National University. Members of this faculty do a minimum amount of teaching, which is largely done by members of the teaching part of the university. The research faculty likewise have a maximum amount of funding to support their research.

Figure 6-3. Professor Francis Lions (1901-72), of the University of Sydney, had obtained his degree in organic chemistry. However, he is best known for his syntheses of complicated organic chelating ligands in collaboration with the research of his colleague Dwyer.

V. Gutman

Professor Viktor Gutman (**Fig. 6-4**), of the University of Austria, and I became friends when I chaired a session at the ICCC. He and Professor Russell Drago almost got into a shouting disagreement on their differences on some acid-base concept. Russ, then from the UI, was known to get over-involved in discussions. I was able to calm him down and move on to the next speaker.

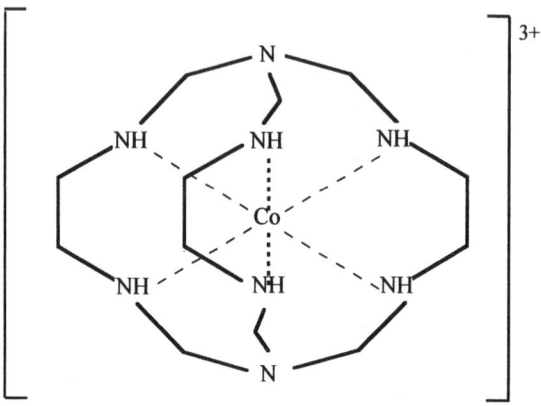

28

Figure 6-4. Professor Viktor Gutman (1921-98), of the University of Vienna, was an expert on acid-base reactions in nonaqueous solvents. He is best known for his book on the subject.

When at NU, Viktor was a guest at our house, and he reminded me about the incident with Drago. In return, when I was in Vienna, he had only two tickets to the opera, and he gave his tickets to me, insisting I take his beautiful young, blond wife to the opera. I believe I won on that exchange.

S. Asperger

Professor S. Asperger (**Fig. 6-5**), of the University of Zagreb, spent one month at Northwestern, teaching part of my course on coordination chemistry. He had worked with Sir Christopher Ingold on kinetics and mechanisms, and was therefore interested in this area of our research. He played tennis, and seemed to like beautiful women who smiled at him. He hosted me when I was in Zagreb at the invitation of the Croatian Academy of Sciences. While there, I heard the Croatians make some snide remarks about the Serbs. My next stop was Belgrade, at the invitation of the Serbian Academy of Sciences. There I heard the Serbs make similar remarks about the Croatians. Little did I ever expect this would escalate into a savage war.

A. A. Vlček

Dr. A. A. Vlček (**Fig. 6-6**), of the Institute of Physical Chemistry and Electrochemistry in Prague, Czechoslovakia (now Czech Republic), obtained a National Science Foundation Senior Foreign Scientists Fellowship to come spend ten months in our department, starting in October 1964. The Institute, prior to the Russian occupation, was named the J. Heyrovsky (Nobel Prize 1959) Polarographic Institute, after his discovery and development of this method of analysis.

There was considerable correspondence between Tony Vlček and our Department Chairman, Professor Donald DeFord, and with me, before arrangements were made for Tony and his family to come to NU.

Figure 6-5. Professor Smiljko Asperger, of the University of Zagreb, got his degree with Sir Christopher Ingold. Asperger's research was on the kinetics and mechanisms of metal complexes.

Figure 6-6. Professor Antonin A. Vlček (1927-99), and his wife. He was director of the J. Heyrovsky (Nobel Prize, 1959) Institute of Physical Chemistry and Electrochemistry in Prague, Czechoslovakia.

This was at a time when Russia allowed precious few scientists to go to Western countries, not even to scientific international conferences. Surely they would not permit an entire family to go to the United States. Tony was presented with permission to visit the US, with the restriction that he could come alone or with his wife, but not with his son. The reason given was that Tony, Jr., age 9, would lose one year of Czech school. Since Tony's wife was a school teacher, they were finally able to convince the authorities that she would teach him his class work during the year, and that he would thereby be prepared to enter the next year's class when they returned. This was approved, and the Vlčeks spent a pleasant year with us.

Tony was an outstanding chemist. Being in the Polarographic Institute, he was an expert on polarography, as was my colleague, Professor Donald Smith (1936-85), who did some research of interest to both. Tony had the same interaction with our research group. He was a versatile chemist doing mostly analytical and inorganic chemistry, while having a good grasp of theory. He gave a few lectures in some of our graduate courses, and he also presented a couple of seminars.

At our faculty conference room brown bag lunches held every day, we talk about any subject we want, at times even chemistry. At one of these luncheons, the topic got around to politics and, at some point, one of us, forgetting Tony was present, said, "Oh, don't worry about it, he is just a Communist." We then apologized to Vlček who said, "That's OK, we would say, 'Oh, don't worry about it, he is just a capitalist.'" I did confidentially ask Tony if he was a member of the Communist Party, because I thought he must be since the Russians had appointed him as Director of an important institute and because he was permitted to come to the US with his family. He said that he was not a member, but that a few such appointments were made to show that the occupation did not mean that all significant positions were filled only by members of the party.

One final comment. The nine-year-old son, Tony, is now in his late thirties. Dr. Antonin Vlček, Jr., has followed in his father's footsteps, and is a distinguished chemist.

C. E. Schäffer

Claus E. Schäffer is Professor at the University of Copenhagen, Denmark and also Chairman of the Chemistry Department. I met him when on our sabbatical in Copenhagen. At that time, he was a young student of Professor Jannik Bjerrum. Claus was doing research on the syntheses and characterization of chromium complexes. Because the work he was engaged in was close to some aspects of our research, he and I became good friends and we had some interesting discussions about our research. A few years ago he, his wife, and daughter came to the UI, where his wife was collaborating on research with a member of the faculty. My wife and I were very pleased to have them come visit us for a few days. I was particularly glad to hear Claus' seminar to learn about his recent research. During the years since our early acquaintance, he had become interested in theoretical work and his seminar was entitled, "Oxidation States and d^9 Configurations in Inorganic Chemistry." Unfortunately, I was really not able to understand much of his lecture, but we continue to remain good friends.

C. K. Jørgensen

Klixbüll Jørgensen was a Professor at the University of Geneva, France. As mentioned earlier, he is perhaps a genius, because many of us chemists are not able to really understand his chemistry. He visited us twice and gave seminars each time. The first time he spoke mostly about the chemistry of the rare earth elements and the actinides. Several years later his talk was more related to physics than to chemistry. When here at NU, he spent more time with Pearson than with me. He died recently after a long illness.

M. Becke-Goehring

Margot Becke-Goehring (**Fig. 6-7**) was a Professor at the University of Heidelberg, Germany. She was one of the very few visitors whose research and seminar was not on metal complexes. Her research was on main group chemistry, specifically, the syntheses of new compounds containing H, B, C, N, and S. Her research group had prepared a variety of compounds, some of which may detonate when struck with a hammer. Her seminar consisted of slide after slide, showing more and more of the compounds made in her lab. Later one of my colleagues, Robert Burwell, said, "We could do without speakers who insist on telling us about the syntheses of each and every analogous compound prepared in their lab." After Margot retired from the university, she became Editor of the *Gmelin's Handbuch der Anorganischen Chemie*.

Figure 6-7. Professor Margot Becke-Goehring, of the University of Heidelberg, was known for her research on the main group elements. Later, she became Editor of *Gmelin's Handbuch der Anorganischen Chemie*, which was widely used by inorganic chemists of my generation.

I first met Margot when Mary and I were driving from Copenhagen to Rome in 1955. We arrived at her office with no prior notice of our coming. She was housed in an old building that had been used by the late Professor R. W. Bunsen (1811-99) and his bust outside could be seen from a window in her office. He was noted for his chemistry, but became best known for devising (1855) the gas burner that bears his name, and which was used for many years in chemistry labs when heat was required. Margot said she was tired of using Bunsen's old desk and lab, even if it was of historical interest. Nevertheless, I recall her saying that good chemistry can still be done in old labs. However, she asked her assistant to drive me out to see where the new science building was being constructed that would house chemistry. Margot was one of a relatively small number of women doing good research in inorganic chemistry—it has taken many years but there is now a large number of women chemists.

E. O. Fischer

Professor E. O. Fischer (Nobel Prize 1973) at the TUM and I have long been friends, since 1955 (**Figs. 2-10, 5-13, 5-14**). He wanted his friends to call him E.O., which I do. I believe he has never published a scientific paper in English, because he says the German language allows one to express things more clearly and meaningfully. On one of his earliest visits to the US, he was our house guest. I told him about the elderly German lady who lived next door to us who wanted to meet him. I introduced her to him, and then left the two by themselves. After some time, E.O. returned to tell us how he enjoyed having her tell him about her childhood in Germany. However, he did say that he had some difficulty understanding her German. A few years later I was a house guest of his. He was not married and lived in his father's house, being cared for by his father's maid who took care of E.O. as a child. She continued to do everything for him in order to be certain that he was comfortable. She did finally die, and E.O. got married at a rather old age. He and his wife are very happy so he now has little time for chemistry.

Fischer's spoken English was adequate, but at times he found it difficult to find a correct word. On one of his visits with me at NU, he was on his way to give the opening talk at an International Conference. He said he had practiced his talk and believed it to be OK, except for one thing. This was the phrase, "one electron not enough." Although he realized chemists would understand it, he wanted me to help him with the phrase. I told him Americans would say "one electron deficient" or else "deficient one electron." He has long remembered this and still thanks me for having helped him.

One other thing about E.O. that is of interest to experimental chemists like me: If one only counts electrons, then it would seem to follow that, like ferrocene, there should be a stable $Cr(C_6H_6)_2$ (dibenzene chromium). An in-depth understanding of the bonding in this compound requires its interpretation by a theoretical chemist. Professor Wilkinson, at Harvard, asked his theoretical colleague if he could make a calculation to see if $Cr(C_6H_6)_2$ would be stable. After a few days, his colleague returned to say, "No way this compound could be stable. Do not waste the time of your people trying to make it."

Some weeks later Wilkinson read that Fischer had made the stable $M(C_6H_6)_2$ compounds where M = Cr, Mo, or W. Wilkinson confronted his colleague with this, and he said, "...it must be wrong, but I will check my calculation." The next day he returned to say, "Oh yes, these compounds should be stable. I had overlooked one of the necessary parameters in my equations." I was glad to have Fischer tell me this, because it shows that the pedestrian counting of electrons, which even I understand, sometimes works.

W. Klemm

During the time that Ron Nyholm (**Fig. 2-19**) was with us, one of our guest seminar speakers was Herr Professor Dr. Wilhelm Klemm (**Fig. 2-12**) of the University of Munster. He was a member of the old generation, when students looked upon Professors as "Gods" and some of them began to think it was so. One of the stops Mary and I made on our drive to Rome was at the University of Munster, where I could meet Klemm and have him tell me about his research. He was not there, so his assistant kindly showed me their labs and told me briefly about the nature of their research. Later, when Klemm came to visit us and I was driving him to NU from Midway airport (there was then no O'Hare airport), I told him that, as he was not there when I visited his Institute, his assistant was very helpful, showing me the labs and something of their campus. He said, "Yes, my assistant is the 'best sheep in my herd.'" That remark shocked me, and I felt like letting him off to have to walk the rest of the way to our department.

That evening I hosted him and Nyholm for dinner. In those days Evanston was "dry"; no alcohol was permitted by local option. We generally would take a guest to our house and offer them a drink, while discussing where to go for dinner. When I asked Klemm and Nyholm what they wanted to drink, they said, "The same as you." I replied that I would have a dry martini, and they volunteered that they liked martinis. What they did not know is that our martinis are mostly gin with a dash of martini wine, whereas the Europeans use just the opposite ratio. After two American martinis, I took two bottles of good red wine and we went to Fannies for an Italian meal. This was the only restaurant in Evanston where one was allowed to bring his own bottle of wine. I often went there with guests, and I knew Fanny, for we always talked a little Italian. After a good meal and two bottles of good red wine, I suggested that we go to a bar in Chicago for a beer. We did, and the ladies appeared one at a time to do their dance and striptease. Once we saw the acts and finished our beer, I would always suggest we leave, and we did. This time, Klemm said, "Oh no, it is my turn to buy the beer and to see again the dance." I explained to him that it would be the same ladies in (and out of) different costumes. He insisted, so we stayed. Later I met one of his assistants at some meeting and told him about my evening with Klemm, and the assistant said, "Oh no, that just could not have been Professor Dr. Klemm."

V. Balzani

Vincenzo Balzani is Professor at the University of Bologna in Italy. He visited us only once and gave a truly fantastic lecture that was of extreme interest to both inorganic and physical chemists. He presented his research on the synthesis of what I believe to be the first dendrimer, making use of 22 ruthenium (Ru) metals holding together the organic ligands (**29**). The inorganic chemists were interested in the strategy used to prepare such a large metal complex. The physical chemists were interested in the spectroscopic and electronic properties of the compound. Vincenzo is known worldwide for his high quality of research and for his choice of significantly important problems. At a relatively early age, he was elected a Correspondent Member of the Italian Academy of Sciences, Lincei.

29

2,2'-bipyridine (bpy)

2,3-bis(2-pyridyl)-pyrazine (2,3-dpp)

Dendritic complex is held together by 22 ruthenium ions.

I. Bertini

Ivano Bertini is Professor at the University of Florence, Italy. He obtained his Ph.D. with Sacconi as his mentor, and Ivano is following in the large footsteps of Sacconi. Ivano has visited us several times, because he and my colleague Tom O'Halloran are involved in some collaborative research, and because he and I are the best of friends.

He is surely one of the experts in making good use of paramagnetic NMR to help elucidate significant biological structures and possible reactions. He has the only 800 megahertz NMR in Italy, partly funded by the European Union (EU). His recent seminar here showed how powerful these NMR spectra can be when in knowledgeable hands.

Ivano is an outstanding bioinorganic chemist, and a gregarious extrovert. His science and his presence are always noted at any scientific conference he attends. This expertise in bioinorganic chemistry, coupled with his energetic enthusiasm, resulted in his arranging the first International Conference on Bioinorganic Chemistry in Florence. This conference set the stage for an upsurge of conferences and journals in bioinorganic chemistry. With all of these attributes, it came as no surprise to learn that he has now been elected a Correspondent Member of the Italian National Academy of Science, Lincei.

F. Calderazzo

Fausto Calderazzo (**Fig. 3-20**), Professor at the University of Pisa, Italy, has been our guest a couple of times. He is a chemist who does very careful work, often on new inorganic systems. One can never really be certain what is to be the topic of his seminars. For the excellence of his research, a few years ago he was elected Correspondent Member of the prestigious Italian Academy of Sciences, Lincei. Fausto is a quiet, private person, but when known better, is a charming person to have as a friend.

A. Ceccon

Alberto Ceccon is a Professor at the University of Padova, Italy. He is a physical chemist doing research on organometallic systems, and I think of him as a physical inorganic chemist and a friend. He became interested in some of our research on ring slippage (see pg. 106) in nucleophilic displacement reactions of cyclopentadienyl or indenyl metal complexes. In fact, he made a more complete study than we on systems related to our research. Our research group was pleased to find that he was easy to interact with, and eager to discuss areas of mutual interest in ring-slippage.

I. L. Fragala

Professor Ignazio Fragala, of the University of Catania in Sicily, has visited our department several times, because he and Tobin Marks, in our department, collaborate on joint research projects. Ignazio does theoretical work on d- and f-element organometallic compounds, while Tobin does experimental work on the same compounds. This means that Ignazio and his research group have made seminal contributions to understanding structure and bonding in group 4 metallocene, organolanthanide, and organoactinide complexes. He has given us seminars on some of his visits, but these were on his research, which involved too much theoretical chemistry for me to understand.

Tobin is his host, but my family and that of Fragala's came to know each other when they spent the summer here so he could work closer with the Marks research group. During the summer, they visited us at our home, and they were interested in seeing my vegetable garden. Fortunately, it was the time to pick the tomatoes and peppers, so they were given some and later would tell us how much they enjoyed our cook-out and large back yard with garden.

I have gone to Catania twice, and each time I made certain to spend some time with Ignazio and his family. They have always been very generous in taking us to see some of the sights in and near Catania.

R. Romeo

Raffaello Romeo is Professor at the University of Messina, Italy. I first met Raffaello years ago when invited to give a lecture at his university. He was waiting to see me to tell me about his research, and to express his frustration with not having his papers accepted for publication. He said that the reviewers always managed to find some trivial thing that he could take care of easily. However, the one problem that he was having had to do with the chapter in our book on ligand substitution reactions of platinum. At the time we wrote our book, the literature on this subject reported that Pt(II) complexes react by an S_N2 mechanism. There seemed to be good reason for this, and we may have overreacted by implying that all such reactions go only by this mechanism. Raffaello had observed what he said to be S_N1 reactions for certain Pt(II) complexes. He said that the referees of his papers would point to the Basolo-Pearson book which "says" that these reactions will "always" be S_N2. According to him, this was the primary reason why his papers were not accepted for publication. I asked him to give me a copy of the manuscript, so I could study it. I did read it carefully, and his kinetic data strongly supported an S_N1 reaction. Perhaps the reason for this change in mechanism is caused by there being one or two methyl groups on Pt(II). Since this results in a higher electron density on Pt(II), it means that the Pt-L bonds are weakened enough to cause a bond breaking S_N1 reaction. I did discuss Raffaello's problem with the referees, and he was soon able to get his papers published.

Raffaello's seminar was on this research, and it was quite clear that the Pt(II) complexes that he was using do indeed involve S_N1 reactions.

D. A. Buckingham

David Buckingham, Professor at the University of Ortago in New Zealand, was also a student of Frank Dwyer. Dave was initially at the Australian National University in Canberra. His research was concerned with attempts to use metal complexes to produce desired polypeptides. Dave and Sargeson coauthored a few scientific papers. Dave was in the US on a lecture tour, and one of his stops was at the UI to be honored by receiving the Bailar Medal. Since Dave was to be so near us at NU, I invited him to come visit and give a seminar. His seminar was on the research mentioned above.

Dave did bring me some personal news that caught me by surprise. Bailar's wife, Florence, had died a few months earlier. My wife and I stopped at Bailar's house to give him our condolences before going to the church for her memorial service. When John saw me he almost burst into tears, for he and Florence were always together and so much in love. Dave came with the news that John had just gotten married. He told me that John had gone to his alumni reunion where he met his high school girlfriend, who had recently lost her husband. The two of them renewed their acquaintance, and decided to marry. She was considerably different from Florence, but I think she made John happy in his later years.

P. Sobota

Professor Piotr Sobota, of the University of Wroclaw, Poland, obtained his Ph.D. degree with Professor Trzebiatowska (see below). Piotr and I met in 1993 when he invited me to come to Poland to participate in the *Summer Schools on Coordination Chemistry* that he and Professor Jozel-Ziolkowski had organized. These schools, to be held every three years, were mostly for the benefit of their students, who would not only learn some coordination chemistry, but also would meet some of the senior chemists doing work in the area. This was a three-day meeting with about fifteen of us giving talks and having ample time to interact with faculty and students. The meeting was held in a beautiful mountainous region of southwestern Poland. The meeting, as well as our housing, was in a suitable ski cottage. I am told that these summer schools continue, and that they are very helpful to their faculty and students.

Piotr's research is concerned with the role of magnesium (Mg) and manganese (Mn) salts as catalyst components in Ziegler-Natta polymerization by the early transition-metal compounds. His work involves isolation of various solid polynuclear metal-containing species from catalyst mixtures, determining their structures by X-ray crystallography and relating their structures to the activity in polymerization. The seminar on his research came to most of us as something completely new and different. Although thousands of articles had been published on research done globally to better understand the catalytic details of this polymerization, none to our knowledge had used the Piotr approach, which helped to make his seminar a big surprise and of interest to us.

After his seminar, he and I had a pleasant dinner together. We talked more about our personal lives than chemistry. This meant more about our families and even more about life in Poland which had suffered during the years of WWII and then by the occupation of their country. On a lighter note, I recently visited with Piotr at the celebration of the 65th birthday of my Ph.D. student Andrew Wojcicke, who was also born in Poland and is a good friend of Piotr.

B. J. Trzebiatowska

Professor Boguslawa Jezowska Trzebiatowska (**Fig. 3-14**), at the University of Wroclaw, Poland, and her husband belonged to the Polish Academy of Science. He had been President of the Academy, and also a visiting scholar doing research with Professor Selwood in our department. She was a very serious and intense person who was therefore hard to deal with at times. Since she and I attended many ICCC meetings, we came to know each other rather well. Most of our conversations were about research being done in our respective labs. She always had a large and able research group (see Sabota above). Because of her large research group, she had on-going work in several areas of research and, therefore, at the ICCC she would present one or more talks on her work. We began to have a problem with her because she would not keep to the time allotted for her scheduled talk. Chairmen of the sessions tried desperately to have her stop, but she would ignore them and continue on for an additional 10 to 20 minutes. This meant speakers who followed her were not on schedule and would have to give up some of their time. The problem was solved when she was scheduled as the last speaker before the lunch break or of the session. Thus, she could have as much time as she desired and the listeners could leave when they wanted.

The madam, as we called her because of the difficulty with pronouncing her name, hosted the 1970 ICCC in Krakow and Zakopane. She arranged a good program, and she invited me to give one of the plenary lectures. I had to decline because I had just taken on the chairmanship of our department and it seemed prudent to stay and get off to a good start in my new assignment. She let me know how she had counted on me and could not believe that I would turn her down. The next time I found it necessary to excuse myself from her request was at the time when I became president of the ACS. One of her research projects dealt with synthetic oxygen carriers, an area in which we too were doing research. She suggested that we collaborate on this and apply for joint funding. I declined because my other commitments were already more than I could handle.

In spite of these two setbacks, we continued to be the best of friends. Therefore, in 1971, it was a sad day for me when I was told that she had fallen down the stairs in her house, not surviving the fall.

M. E. Vol'pin

Professor Mark Vol'pin (**Fig. 6-8**), Director of the A. N. Nesmeyanov Institute of Organoelement Compounds, was clearly one of the very best inorganic chemists in Russia and in the world. He and I became friends in 1971 when I attended the International Conference on Organometallic Chemistry in Moscow. During the week of that meeting, Mark and I were able to talk privately about our research interests and about our lives in general. This was during the period of the cold war between the US and the USSR, when scientists who were not members of the Communist Party were not allowed to travel to western countries, particularly the US. The ICCC tried, more than once, to arrange permission for Mark to give a plenary lecture on his research, but this was always denied by the USSR government.

Mark and I met again when we both attended a scientific conference in Poland or in Czechoslovakia, I cannot now recall which. Both countries were then occupied by the Russians, so their scientists were permitted to travel within the Eastern countries behind the iron curtain. We talked about his research on graphite-transition metal complexes and I came away thinking that he might even be attempting to find some way to make diamond.

The next time we saw each other was in 1993 at NU, when my colleague Tobin Marks and Vol'pin arranged a workshop on Organoelement Chemistry funded by NSF and the Institute of Organoelement Compounds (INEOS). Mark and a small group of Russian chemists came to participate in the workshop. They presented and discussed their research, and we did likewise with ours. This was a very stimulating three days for both groups. Not only did we get to exchange views on our work, but we also had pleasant evenings together—helped along with excellent Russian caviar and vodka.*

Figure 6-8. Me at left with Professor Mark Vol'pin, Director of the A. N. Nesmeyanov Institute of Organoelement Chemistry (1993-96). His research in inorganic chemistry was always carefully designed to answer important questions.

The next time Mark and I were together was in 1994 when he invited me to attend Workshop-INEOS-94 in Moscow-St. Petersburg. This workshop was held on a tour boat, slowly going from Moscow to St. Petersburg. Since we could not get off the boat, we were not distracted by our surroundings as we may have been if it were held in a city. I arrived in Moscow a little late and was tired, so I went directly to my room. The next

---

*Having mentioned caviar, I want to digress on one thing that happened at the conference I attended in Moscow in 1971. Nesmeyanov, who hosted the conference, was the most important chemist in Russia. He was the Director of INEOS and had received the highest honor of a star, awarded to very few citizens of Russia. During that conference, he invited a few of us to visit his Institute. We were driven there in limousines, and the people were prepared to show us their lab and describe a bit of what they were working on. We finally were returned to a conference room next to Nesmeyanov's office, where they had prepared coffee or tea and snacks of various types, including caviar. He shortly appeared to greet us. After some conversation, he asked if we liked the caviar. We said that we found it to be very good but, as we did not often eat it, we were not qualified to judge. He then proudly explained that it was synthetic caviar made in his Institute. He explained how this was accomplished. The globules were of protein, salt was added in the correct amount, and ferric tannate was added for the proper color. The difficult problem that they experienced was to duplicate the odor of caviar. However, by using gas chromatography they found the active compound responsible for the natural odor. They used the compound to complete the synthetic caviar. We left puzzled as to *why* they wanted to do this, since Russia has a resource of sturgeons which provide them and the world with real caviar.

morning I was informed at the reception desk that I had a personal invitation to have dinner at Mark's house. The workshop started with an evening gathering for us to get acquainted. Mark attended and tried to talk with most of the persons present. He looked ill and told me that he would not be able to attend the workshop. He asked me if I would take on his function of ensuring that the talks and discussions were kept on schedule. This was the last time that I saw my good friend and outstanding chemist. It was sad and a great loss when he died on September 28, 1996.

L. G. Sillén

Professor Lars Gunnar Sillén (1916-70; **Fig. 6-9**), of the Royal Institute of Technology (KTH) in Stockholm, started his career in crystallography, doing his Ph.D. dissertation on crystal structures of bismuth (Bi) oxides. He decided it might be more fun doing some other type of research so he moved to the other extreme and spent the remainder of his professional life working on solution chemistry. One reason he moved to solution chemistry was that his department did not have sufficient funds to support salaries and expensive equipment. He said, "There were no salaries for research work; we had to earn our living by hard teaching (sometimes even by teaching organic chemistry)." In spite of such problems, Lars ended up making some important contributions to the equilibria and species of metal ions in solution on the addition of base. He developed graphical methods to make such studies and, in 1960, even developed a computer program to enhance his research. During the latter days of his short (54) life, he was beginning to model the composition of, and the reaction processes in, seawater.

Figure 6-9. Professor Lars Gunnar Sillén (1916-70), of the Royal Institute of Technology (KTH), was well known for his detailed studies of the species of metal ions in water solution versus the solution pH. Shortly before his untimely early death, he had begun to study the nature of metal species from different parts of the oceans.

Lars and I often attended the ICCC where we would visit with each other and talk about our research. I would often tell him that his studies were much more difficult than ours, because one may get hydroxy metal species in solution that cannot be identified by his method of base titration. He had somewhat similar comments to make about some aspects of our research. The ICCC of 1962 was held in Stockholm with Lars as its chairman. What we all remember about that conference is what took place at the morning introductory session when all of the attendees were present. At one point Lars reminded us to keep to the time allotted for our presentations, stating that a yellow light would blink to indicate when the speaker only had two more minutes, followed by a red light which indicated that only one minute was remaining, and, finally, a large strong Swede would then come out to pick up the speaker and carry him off the stage. Since Lars was a large, tall, and strong person, he demonstrated how this would be done by picking up one of the people on stage and carrying him off to a back room. Those of us in attendance always recall this, but it had little effect because some of the speakers still spoke over their time limit.

Lars gave an excellent talk in our department on his research. He started by asking if we would prefer to have him speak in German, Russian, French, Spanish, Japanese, or Latin. Clearly his talent was not limited just to chemistry. After his talk, the two of us and a couple of my colleagues went to my house for a before-dinner drink.

Figure 6-10. Professor Gerhart Schwarzenbach (1904-78), at the ETH, is known best for work on chelating agents leading to the discovery of EDTA used to sequester metal ions in solution. This required an extensive study of the chelating properties of different multidentate compounds designed for different purposes.

Lars did not drink any alcohol, and explained that in Sweden if one was in an auto accident after having too much to drink, his license would be confiscated for life. Thus, if one drinks at some party, it was first decided who would be the sober, designated driver.

It was a very sad day when we coordination chemists learned about the early death of Lars.

G. Schwarzenbach

Gerhart Schwarzenbach (**Fig. 6-10**) was Professor at the ETH in Zurich, Switzerland. He was well known for his careful detailed study of the stabilities of metal complexes in solution. His studies led to the universally used ethylenediaminetetraacetate (EDTA) to sequester metal ions in solution.

Mary and I stopped in Zurich on our way to Rome in 1955. One thing I wanted to do was to meet Gerhart for the first time, and have him tell me about his research in progress. When I arrived at the ETH and asked to see him, I was told that he had just left to do two weeks of military duty. My first reaction was that they were kidding me, because he was already in his late 40s. What I did not know at the time is that Switzerland has a Swiss army which requires all eligible men to spend some time in military training each year, even though Switzerland has not "ever" participated in any war. There is considerable talk, along with voting, on the issue of whether Switzerland should continue to maintain an army.

Our research interests were much the same, so we finally did meet at some scientific conference. Gerhart visited me once and gave a first rate seminar on his research. He took us through his work and indicated how he arrived with EDTA having just the correct geometry to attach itself to a six-coordinate octahedral structure of almost any metal ion. This property of EDTA is why it is so efficient in sequestering any metal ion in solution. For example, it can form a chelate with calcium (Ca) in hard water, preventing it from acting as hard water.

L. Venanzi

Luigi Venanzi (**Fig. 3-18**) was a Professor at the ETH in Zurich, Switzerland. Luigi was graduated from the University of Roma and then he went to Germany (with Klemm) and then directly to Great Britain. Years later, he came to the US, first at SUNY-Albany, then as chairman of the department at the University of Delaware. After that, his next and final move was to the ETH. Luigi's research as a coordination chemist was outstanding, and his lectures were worthy of the "Olympic Gold Medal." He knew three or more different languages. At times, he was an interpreter for me in Germany. He spoke Oxford English even better than do most persons at the University of Oxford. I feel certain he could have returned to Italy at a good university position, but he chose not to, even after his retirement. He and his family remained in Zurich. Chemists worldwide knew him for his chemistry and for his delightful persona. He died in 2000 and will be missed by us, but not forgotten.

Figure 6-11. Professor Yatsimerski was Director of the Institute of Physical Chemistry of the National Academy of Sciences of the Ukraine. His primary research interest was determining the stabilities of metal complexes in solution.

K. B. Yatsimerski

Professor Yatsimerski (**Fig. 6-11**) was the Director of the Institute of Physical Chemistry of the National Academy of Sciences of the Ukraine. During the cold war, Russia did not permit many of its scientists to travel to western countries to attend scientific conferences. Yatsimerski was allowed to come to one of the ICCCs. It was there that we met and struck up a professional and personal friendship. I invited him to come visit us the next time he was in the US. A few years later, he did come and gave a seminar in good English. His research had to do with the measurement and understanding of the stability of metal complexes.

Since his hosts had been taking him to dinner in restaurants, I suggested he may like to come to my home and join my family in a cook-out. He was delighted to be asked, and we were very pleased to have him as our guest. It was a beautiful sunny day, and my children and their friends played basketball on our patio. The teams consisted of girls and boys; they all played with much intensity, wanting to win. Yatsimerski said he had never before had a cook-out, nor seen such a vigorous game played with boys and girls on each team. When he returned to the Ukraine, he wrote to thank me and say that his stay at NU was the highlight of his travel in the US. We have kept in touch all of these years by the exchange of Christmas cards.

C. Addison

Professor Cliff Addison (**Fig. 2-18**), of the University of Nottingham, played an important role in the post WWII development of inorganic chemistry in Britain. He was largely responsible for building up the department of chemistry at Nottingham, assisting it to become internationally renowned for both its excellence in undergraduate teaching and its substantial contributions to research in inorganic chemistry.

When visiting NU to give a talk on his research, he and his wife were our house guests. Since we had been his house guests when visiting them in Nottingham, this gave the four of us a chance to again compare family matters, such as the children now having become young adults.

Cliff's seminar was on chemistry almost unknown to our inorganic faculty, because he was the "only" chemist worldwide investigating the solvent properties of N_2O_4 and its powerful oxidation properties. He described, in some detail, the safety precautions needed when working with N_2O_4. Despite the great care taken, he gave us a few examples of explosions in his lab. This convinced us that we would not want to do research using N_2O_4.

J. Chatt

Professor Joseph Chatt (**Fig. 2-20**), at the University of Sussex, started his career by doing basic research in industry at the Imperial Chemical Industries (ICI). After some fifteen years, he left to become Professor at the University of Sussex. There he had the position of director of the lab on nitrogen fixation. With the help of some very able assistants, such as G. J. Leigh, they did some important pioneering work on attempts to convert nitrogen (N_2) into some of its compounds.

I think the first time Joseph came to the US was in the late 1950s when we invited him to come give a lecture at the GRC. He took that occasion to visit me at NU and give a seminar. In England he is called Joseph—in the US, Joe, as mentioned earlier. I know of no other chemist who conducts his research with greater care. Research scientists say, "If a scientist does not want to make a mistake, albeit minor, he should not do research." To my knowledge, Joe, by his careful research, proved this statement to be false.

The seminar he gave was on some elegant pioneering work on the syntheses, reactions, and properties of some new platinum (Pt) complexes. Joe was very likeable and pleasant, but formal. Therefore, it surprised me when he started his lecture by telling this story:

> On a commuter train to London there were two passengers who caught the train at the same time, and who often sat across from one another. One of the passengers had a pad of paper and kept throwing a sheet out the window during the ride. After some days of seeing this, the other passenger asked, "Why are you doing this?" His answer was "To keep the lions away." The questioner said, "But there are no lions in England," and the response was, "Yes, so you see it works."

Joe wanted to make the point that in basic science it is not enough to know it works, but one also needs to know why it works. He then proceeded to illustrate the point in the talk about his elegant research.

Joe will long be remembered by inorganic chemists for his outstanding research, particularly in platinum chemistry. It's my opinion that he is "the father of platinum chemistry in the UK." I am pleased to see that he is being honored by the Royal Society of Chemistry which is awarding the Chatt Medal to an inorganic chemist each year. (I feel humble but honored to have been the first Awardee.) Joe started the ICCC in 1950, and, in the year 2000, the conference of about 1,500 attendees held in Edinburgh was dedicated to him. Very recently, the new Nitrogen Fixation Laboratory was named Joseph Chatt. That he is highly deserving of these honors is well recognized by inorganic chemists.

H. J. Emeleus

Professor Harry Emeleus (**Fig. 3-5**), at Cambridge University, was an outstanding senior inorganic chemist. His research had to do with main group elements, whereas mine dealt with transition metals. We were both inorganic chemists in its early days and had a general interest in each other's work, but did not dwell on specific details. As mentioned earlier, I had visited him in Cambridge in 1955. I knew he was to be in Chicago so I invited him to come give a seminar. His research then was on reactions in nonaqueous inorganic solvents. That was a time when there was considerable research activity in this area. For example, one of the most studied was liquid ammonia (NH_3). It came as a complete surprise to me that the solvent he was using was bromine trifluoride (BrF_3). This was to be the first such use of this solvent in interesting acid-base reactions. While having dinner together, I recall that he was a bit unhappy because, although he had discovered this system, others had read his papers and, finding them of interest, began to do similar research using "his" solvent. I told him that this was often done by US chemists, even if not by UK chemists. I thanked him for his book with Anderson entitled *Aspects of Modern Inorganic Chemistry*. He was pleased to hear that I was able to make good use of the book to prepare lectures for a course I was teaching at a time when I felt there was no suitable textbook on inorganic chemistry.

J. Lewis

Professor Jack Lewis succeeded Professor Emeleus at the University of Cambridge. Cliff Addison (**Fig. 2-18**, p. 56) was Jack's research advisor and Ron Nyholm was mentor and advisor to Jack regarding academic activities. Jack was at UCL where Ron had brought together several outstanding young chemists, resulting in their lab clearly being the best in the world in coordination chemistry. (This is no longer a primary research interest at the UCL.) I have known Jack for many years, even when he was not happy because, having gotten his degree in a "red brick" school instead of either Oxford or Cambridge (the Harvards of England), he therefore thought that he would never be appointed to some highly responsible position. In spite of this, he received the Emeleus Chair at Cambridge University, was largely responsible for establishing Robinson College, became a Fellow of the Royal Society (FRS), was knighted to Sir Jack and, recently, elevated to "The Lord Lewis of Newnham." He was elected a foreign member of the US NAS in 1987.

Little did we know that we were hosting a future Lord when he twice visited us and gave two first rate seminars. I cannot recall for certain, but I think his first seminar had to do with some of the Nyholm research on nickel complexes with diars (**30**). A few years later, he presented his research on the pyrolysis of osmium pentacarbonyl

benzene ring with $As(CH_3)_2$ and $As(CH_3)_2$ substituents

30

($Os(CO)_5$). This heating of $Os(CO)_5$ caused it to lose CO and form a variety of different metal clusters. It was then necessary to isolate the individual compounds and, using x-rays, determine their structures. It seemed unfortunate to me that the compounds were obtained in such small amounts that there usually was not enough to be able to investigate their chemistry. Nevertheless, much was learned from the thermal process and the nature of the products formed. In addition to the very good chemistry reported, Jack's lectures were eloquently presented. His talks are well prepared, and his delivery is second to none.

R. S. Nyholm

Ronald S. Nyholm (**Fig. 2-19**) was born in Broken Hill, an isolated mining town in a semi-desert, mineral-rich area in New South Wales, Australia. The streets in Broken Hill have names of minerals, metals, and acids (Blend, Cobalt, Chloride, Bromide, Sulfide). Perhaps one reason we were good friends is because we were both born and grew up in small mining towns. Ron went from his mining village to such heights as to become Sir Ronald.

Ron's first time in the US was to give the invited Riley Lectures in 1957 at Notre Dame. After the lectures, he came to our department to give one month of lectures in my course on coordination chemistry. Ron was an outstanding coordination chemist and at that time his research was mostly involved with the ligand called diars and the metal nickel.

There are two things I remember, other than our interest in the same area of chemistry. One had to do with our going to see a Chicago Cubs baseball game in beautiful Wrigley Field where the Cubs lost, as they often do. It was a nice sunny day, contributing to our drinking more beer than was perhaps necessary. I had taken other foreign guests to see their first baseball game, but all that interested them was the beer, and they felt the game was boring. Not so with Ron, who had played county cricket, whereas I had played baseball. To see a sports game that one has never participated in can be boring, but Ron and I got engrossed in comparing the differences and similarities between baseball and cricket. Even when it may seem as if nothing is happening on a baseball field, there are certain strategies taking place. For example, if a batter is a long or a short ball hitter, the outfielders would stand further away or nearer. If there is a man on first who can try to steal second or who may be moved to second by a batter bunting to try to sacrifice him to second, there are different positions played by the infield, etc. It seemed that everything I pointed out to Ron about what was happening, even when it appeared as if nothing was going on, he would counter with it being similar to a certain thing in cricket.

Ron had just the right personality and aggressiveness to make it possible for him to reach the politicians. He was able to tell them of the importance of supporting and funding science, in particular, chemistry. In fact, he did get them to set aside a budget to support coordination chemistry in the UK. It was a sad day when Ron fell asleep while driving and had a head-on collision with a truck, which he did not survive. Had Ron lived longer, he would have done much more toward generating support for the sciences. I learned of Ron's death when I was awakened in the middle of the night by a phone call from Dr. David Keeton, who told me about the accident. At the time, Dave, who had been a graduate student in my research group, was a postdoctorate in Ron's research group. After receiving this call, I first telephoned Bailar to give him the sad news. We then called other friends of Ron, and jointly wired our condolences and prayers to his family.

A. J. Poë

Professor Anthony Poë (**Fig. 6-12**), of the University of Toronto, was kind enough to come and take care of my students during my sabbatical in Rome (1961-62). I had read some of his publications on the kinetics and mechanisms of halide ion (X^-) exchange with platinum complexes of the type $[PtX_6]^{2-}$. These studies were similar to our research on ligand substitution reactions of metal complexes, so I wrote and asked if he would be interested in coming to NU to look after my research group during my absence. He accepted, brought his family, and they all had an enjoyable year. I think Tony would also agree that he had a worthwhile professional year.

My research group at the time had about a dozen graduate students, and it almost never exceeded this number. A few of my students were doing research jointly with Pearson, and Ralph also interacted with these students during the year. This research was primarily involved with studies of ligand substitution reactions of metal complexes.

Figure 6-12. Professor Anthony Poë, of Toronto University, spent the academic year 1961-62 supervising my research group during my sabbatical year in Rome. He has become an expert on the mechanism of the reaction of metal carbonyl clusters.

We had only recently started our detailed kinetic studies on the reactions of metal carbonyls (pg. 100). Tony was primarily responsible for this group of my students. Since research on these compounds had only been started a few years earlier, and since the work was going well, there was a great deal of research activity and enthusiasm among this group of students. The excitement of these students must have gotten Tony interested in the reactions of metal carbonyls because he has since spent his professional career becoming one of the world's best chemists investigating CO substitution reactions of large complicated metal carbonyls. During the year with my research group, Tony was extremely helpful to my students investigating some of the more simple metal carbonyls. We coauthored several articles on this research.

In addition to Tony's help with our metal carbonyl research, he also was helpful with my other students who were working on different research problems. For example, he worked with John Burmeister (**Fig. 3-13**) on his discovery of linkage isomers of the type L_nM-NCS and L_nM-SCN of metal complexes. Yes, we all were very pleased to have Tony come help us for a year and we remain good professional and personal friends.

M. D. Johnson

Dr. Michael Johnson was on the faculty at UCL. His research was on the kinetics and mechanisms of reactions of Cr and Co metal complexes related to biological systems such as vitamin B-12. This work was similar to our research, so there was considerable interest in our group to hear his seminar. Unfortunately, it was scheduled during protests of the Vietnam War by students on university campuses of the US. The day he was to give his talk, protesting students did not permit us to enter our building. We were about to cancel his seminar but, as it was a nice spring day, I decided that we should go to my house and hold his talk on my patio. There we were able to serve beer and soft drinks. All of us agreed that this was our best seminar ever—but it has yet to be repeated. Some years ago the UK did some downsizing in its universities by giving faculty a choice to retire with a good pension. Several chemists took this option, but kept doing research and part-time teaching. Michael, although young, took the pension and gave up chemistry.

* * * * *

During my 54 years at Northwestern (the last ten as Emeritus Professor) we have always had many foreign visitors of inorganic chemistry. This was true even in the late forties to the mid-sixties, when there was little traveling by chemists. With funding for research and travel expenses, travel has now become commonplace. There are at least two reasons we have always had foreign inorganic chemists visit our department. One is that our inorganic chemistry group at NU has always been at the forefront. This has been ever increasingly true—our group was recently ranked as one of the top four groups in the US (with the University of California-Berkeley, Caltech, and MIT). We are often told that our group is one of the very best in the world. The other reason for our large number of visitors and seminar speakers is that foreign travelers often have to make air line connections in Chicago. This makes it easy for them to stop and arrange a visit with us.

7

EMERITUS PROFESSOR

In 1990, I became the Charles E. and Emma H. Morrison Emeritus Professor at NU. Mary, the children (then adults with their own families), and I looked forward to this day. Of course, this meant that I would be free of teaching and research at NU. Thus, our family could spend more time together, something I had not done earlier because of my travels here and abroad to give invited lectures at universities or at international scientific conferences or symposia. Add to that the years I was President of the ACS and, as described earlier, had to travel worldwide as its representative. Furthermore, my seven years on a NATO science research panel required that I go twice a year to the NATO headquarters in Brussels, and one time a year to a NATO science developing country (see *pp. 148-152*). There were times when I felt as though I was spending all my time in airports, due to delayed or canceled flights. Once I told this to my brother, Martin, who owned the only grocery store in Coello, named "Basolo's Groceries" (**Fig. 7-1**). Martin, who had never left Coello nor had ever seen an ocean or a mountain, said to me, "You are crazy running all over the world, when you could have stayed here and taken it easy." There were times when I was waiting in an airport that I would think about what my brother had said. Fortunately, I did not take his advice, because I may have been a coal miner dying at an early age of "black lung."

Mary and I had prepared for when I would become Emeritus. We lived for years in what we considered an ideal house with a superb location. The two-story house, giving us plenty of room for our family of six, was adjacent to a nice, well kept city park. Mary took advantage of this by using the sand box and swings when the children were young. As they got older, they used the tennis courts and the golf course as well. Mary also appreciated the tree-lined sidewalks and our location on a dead-end street.

She and her friends would often take walks and meet to talk. Mary always enjoyed talking to people, and they to her.

For me, the location of the house was also excellent. It was only a 15 minute walk to the NU campus and my office. On very windy and/or very cold days, I could even make it in 10 minutes! Since I could walk to work, I would leave Mary the car, so we were always a one car family. The location was also ideal when I began to do a lot of traveling. O'Hare airport was about a 40 minute drive from our house. If I were to be gone for only a few days, I would drive and park. Upon my return, the car was there and I was ready to go home. Another advantage of using O'Hare Airport is that when you arrive, you are home, and need not make connecting flights elsewhere. Furthermore, one can fly from O'Hare to any place in the US, regardless how small the town. There may only be one flight a day, but if needed, one can usually arrange to take it.

One thing I did not like about our house was its small lot with many large trees. This meant there was very little sunlight in the backyard, where I would have liked to have planted a vegetable garden. As described earlier, as a child, I helped my father with his gardens. Now as an adult, I would have enjoyed a sufficiently large garden in the sun so I could plant and take care of a variety of vegetables and watch them grow. If fortunate, I might even, at times, have the joy of eating a delicious vine-ripened tomato.

Figure 7-1. Me in front of my brother Martin's grocery store. He and his wife, Amelia, had inherited it from her family. It passed on to my brother's sons, but now is closed due to competition from supermarkets.

Once our four children married and left, the two-story house became much too large for Mary and me. She wanted a one floor ranch style house, and I wanted a large sunny backyard, plus a golf course near by. Mary discussed this with a good friend of hers, Marie Queenan, who lived near us, and worked for a real estate firm. They then began to look for a house that would fit our desires. Both Mary and I agreed that we wanted the transaction to involve selling our house for about the same amount as we would pay for the new house. I asked Mary to look at houses in this price range and when she found one she liked, I would go see it and the two of us could then decide. We looked at a few houses over a period of about a month, and always found something we did not like. Finally, one day Mary had found just the house she was looking for, and said it also had a large, sunny backyard. She called the lady of the house and told her we both were coming to see it.

It was a beautiful sunny day and, when we arrived, the lady was out front waiting for us in her driveway. She and Mary went in the house, thinking I would follow them. Instead, I walked to the backyard and was happy to see it was just what I wanted. Mary had seen the house the first time when there with Marie, but the lady was again showing Mary some of the things she specifically wanted to know about inside the house. When asked, I told Mary that I was sold on the backyard so, if she was sold on the house, we should buy it. We did, and we moved there in 1987. I had previously played golf several times on the Glenview golf course, just a 10 minute drive from the house. Mary got the small ranch style house she wanted, and I got the garden space and nearby golf course that I had wanted. We both were very pleased, and we began looking ahead to 1990, when I became Emeritus.

I had already stopped taking graduate students in 1987, because I felt one's mentor should be available for several years after his students have received their Ph.D. My last student was to have been Dave Kershner. I did make an exception when my good friend Professor Q. Z. Shi of Lanzhou, PRC, pleaded with me to accept his very best student, Jian-Kun Shen, who insisted he had to work with me. Jian-Kun, then, became my last student and, as with Shi, one of my very best students. He obtained his Ph.D. at a time when the job market was very low for Ph.D. chemists. Since he was not able to find employment, I had him stay for two years as a postdoctorate, to give him time to get his green immigration card. This would make it easier for him to get a job. Near the end of his second year, I said to him, "Now we must find you employment." I telephoned a few of my good research chemist friends in industry, and told them that Jian-Kun was outstanding and had done enough work for us to publish 21 papers. In every case, my friends said, "With your highest recommendation, we would hire him in a minute, but we are cutting back on our long range basic research and do not have any openings for Ph.D. chemists."

Sometimes it helps to be lucky, and to attend scientific conferences. Each year, the Martin Kilpatrick lectures are presented at the Illinois Institute of Technology (IIT). Jian-Kun and I attended the two days of lectures. During the first day, I told a research chemist, working at Nalco Chemical, only a few miles west of Chicago, about Jian-Kun. I told him that not only was Jian-Kun an outstanding chemist, but he, his wife, and two boys were very fine people. He was certainly the type person I would want as a coworker. Since everything went so well with this Nalco employee, I told Jian-Kun to talk with him on the last day of lectures, telling him a bit about his dissertation research and asking about the type of research being done at Nalco. Before the day was over, I was privately told by our friend that Nalco had no openings for a research chemist, but that there was soon to be a retirement and he would see if there could be a replacement. Sure enough, as luck would have it, in a month or two Jian-Kun received an invitation to give a talk on his dissertation research at Nalco and to spend the day with them to meet some of their research chemists. I told him not to worry, because I had heard him give a departmental seminar on his research and this would go very well. I further made the point that even more important was that he ask about their research and show a real interest in it. One week later, we were both excited and happy that he received a letter offering him a position. He was to spend a couple of months in their labs learning about the products they manufactured, and about their research. His permanent position would be in a small product control lab in Singapore. He and his wife, particularly she, were very unhappy about this. The point of contention was that they wanted to stay in the US and become American citizens. They knew that jobs here were almost nonexistent, and I told them Jian-Kun would do such a good job that someday he would be transferred back here to work in their central lab. They did go to Singapore, where he now is director of the lab of some twenty people. He often has to come here to the home office to attend some company meeting. Each time he comes, he takes me to dinner and we have a pleasant evening together. A few months ago, he arrived at my house with his wife, two boys, and a bottle of expensive Barolo wine. He and his wife were extremely happy to tell me that I had been correct and that they are now US citizens. His directorship means a much higher salary, an expensive new car, and the prestigious recognition of one in such an exalted position in Singapore. This was a very good time to stop having graduate students—"when you are ahead."

By 1990, Mary and I had been in our new house for three years, and the time had come for me to become the Charles E. and Emma H. Morrison Emeritus Professor of Chemistry. We both agreed that life had been good to us, and we would not have wanted to have it in any other way. Our children were happily married, and we were to have eleven lovely grandchildren. We love and are proud of all of them. None became chemists, but all are teachers with tenure. Earlier I have mentioned several times how I enjoyed teaching, having even received two of the highest ACS teaching awards (see Appendix). Therefore, our children did follow in one half of their dad's footsteps.

The first noticeable happening was in August 17-18, 1990, when there was a large, successful celebration of my 70th birthday. This was arranged by my former students and postdoctorates. For the most part, the gathering was initiated by Harry Gray with most of the work, particularly the local work, being done by my colleague and friend Tom O'Halloran. The complicated work of on-site registration and room assignment was that of my former student Tom Weaver. The design of the Basolo logo (**Fig. 7-2**) was done by some of my professional grandchildren at Caltech, and the T-shirts with the logo were printed by Eric Voss, one of Du Shriver's graduate students. O'Halloran was

Figure 7-2. Youngsters who are my distant relatives living in Milan, Italy (see **Fig. 5-6**). This was in 1990, when I became Emeritus Professor and gave them the T-shirts with my 70th birthday logo. Left to right, the Sgarbi brothers are Dario, Marco, and Fabbrizio.

fortunate in having obtained the use of the Allen Center for all of the activities, including some housing rooms. The Center is NU's Kellogg School of Business building, and it is almost always being used by different groups in the US and abroad who come to attend 1 or 2 weeks of classes and who know that the School has often ranked number one among business schools in the US. The Center is an attractive new building on the lake front of our beautiful campus.

The weather was excellent for the two days of the gathering. Almost all of my Ph.D. students and postdoctorates attended, some even from distant countries such as Australia and China. There were about 60 attendees during the lectures and about 150 at the final reception and dinner. The very informal lectures were given by former students on their current research. The speakers were introduced by postdoctorates who had some anecdotal story to tell that tended to roast the speaker, and often me. The same was true of the speaker who gave informal lectures with several interruptions from the attendees. None of this horse-play was offensive, because we were just a happy professional family having fun. One session was devoted to our "worldwide famous BIP." Blackboard (no slides) lectures on work in progress were given by two current graduate students on their thesis research (**Fig. 3-10**).

Students and postdoctorates were asked to bring pictures they took when here that they could display on the poster boards in place for this purpose. During the reception, most of the people walked around to view the posters. This, too, was a big success, because some of the posters were intended to be amusing, whereas others were serious.

In addition to persons involved during the 1½ days of lectures, some of my best personal/professional friends had kindly managed to attend. Furthermore, my departmental colleagues, the NU President, the Provost, and the Dean of the College of Arts and Sciences were present. After dinner, we were all there for a rather long noisy evening, because many of the people wanted a piece of the action—including the master of ceremony, Harry Gray. The talk designed to roast me was given by my former faculty colleague, Ralph Pearson (**Fig. 3-6**). He knew me well, for we had worked together and published 60 coauthored scientific papers on research done by our students and postdoctorates. Ralph had stories to tell that were true, but also several that he prepared for the occasion. For example, he said when he died and was being shown around by St. Peter, Ralph saw a handsome young man walking hand-in-hand with a very ugly old woman. He asked why the man was walking with this ugly lady, and St. Peter told him that this was the penance he had to do for the sins he did on earth. Ralph then saw me walking with Sophia Loren and asked why Fred was paired with such a beautiful lady, for he knew that I had often sinned on earth. The reply was, "Oh no, he is the penance for Sophia." Needless to say, this resulted in a loud laughter, and it gives a good example of the nature of our friendly celebration.

There were, as well, serious moments after dinner. Harry made the very correct point that had it not been for my dear wife Mary, we would not be having such an occasion. He presented her with a very nice gift (earrings) from all of my graduate students and postdoctorates. He also thanked them, and other friends, for their contributions to the event. Several corporations were also thanked for their support of Basolo 70.[*] As a result of all this generosity, there were enough funds left that my students and postdoctorates decided to work toward an endowment to establish an annual award of a Basolo Medal and Lecture. The endowment has now reached a point where it will support the Award to perpetuity. This is not to say that additional funds are not needed, because additional annual costs for the Award seem to increase each year. I have absolutely nothing to do with the selection of Awardees. This is all done by the inorganic chemistry faculty at NU. The Award is for outstanding research in inorganic chemistry, and it is clear that this goal has been, and is being, reached.[**]

One of the people very much involved with this event was my former student John Burmeister (**Fig. 3-13**). He and Barry Lever, Editor of *Coordination Chemical Reviews*, arranged to have a special issue of the journal dedicated to my 70th birthday. John had told all the speakers to arrive with a manuscript of their talk, otherwise they would not be allowed to speak. This seemed to have the effect that John had anticipated, because they all complied, except one. This exception was Andy Wojcicki who has always been

---

[*] These corporations were: Akzo, American Chemical Society, Amoco, Data Trace, Dow, DuPont, Eastman Kodak, Gordon and Breach, Hoechst/Celanese, Monsanto, Rohm and Haas, Shell Development, Sun, Ube Industries (America), Union Carbide, and John Wiley & Sons.

[**] Awardees of the Basolo Medal: Ralph G. Pearson (1991); Henry Taube (1992); Jack Halpern (1993); Harry Gray (1994); Lawrence Dahl (1995); Richard H. Holm (1996); Kenneth N. Raymond (1997); Malcolm Green (1998); Thomas J. Meyer (1999); James P. Collman (2000); and M. Frederick Hawthorne (2001).

late when dealing with deadlines. However, he was so embarrassed by being the only person who would hold up the publication of the special issue that he promptly wrote his paper and sent it to John. John was then able to get all of the manuscripts and quickly sent them to Barry. Due to Burmeister, this operation set a record that may never be broken. Usually special issues of journals take an inordinate length of time to appear because there are always two or more authors who do not deliver their papers on time. Not only did John take charge and get the job done ahead of schedule, but he also wrote an introductory account of me. He did this so eloquently that he was asked to write the introduction on the occasion of my 75th birthday for *Inorganica Chimica Acta*, and, again, on my 80th birthday for *Chemtracks* (*Inorganic Chemistry*). Unbelievably, John was able to write these three introductions about me, presenting his same message each time, but so done that no two articles were the same.

Mary and I began to enjoy our house, and it seemed we had made the right choice for our retirement. Mary loved everything about the house and its location, except for one thing. She missed her friends on Colfax, and the times they would meet while being outdoors walking on the sidewalk along our street. Our new house in Glenview is in a nice residential area with no sidewalks. This, compounded with the houses being some distance apart, meant that one gets to know only their nearest neighbors. Even this did not result in the close friendships that Mary had formed steadily with the ladies on our Colfax Street. Furthermore, most of the people in the near-by houses are young with both spouses working, leaving little time for anything other than their own families.

One thing Mary and I both enjoyed was having our daughters Peg and Liz living locally with their families. Peg and Gary Silkaitis have two boys, whereas Liz and Bob Pionke have two girls and, God willing in August, two boys. We were so very happy to watch them grow from babies to teenagers. During all of the years, we often were visited each week by them, and enjoyed their calling us Grandma and Grandpa. It was Grandma with whom they were the most comfortable, and when Mary and I were both present, it was clear they always preferred being with her. I would think that the reason for this was that she never failed to give them good things, such as candy, ice cream, and toys. Another reason was that when Peg or Liz were busy, they often would take the children to Mary for care. Since Peg and Liz both teach in an elementary school in Evanston, they had the three months of summer free. This meant they could even spend more time with us, so we were not alone. In addition, MC (Kunzer) and Fred live only about a three hour drive from us. MC and Fred are both teachers, and they likewise have "free" summers. This made it possible for our children and grandchildren (eleven, 6 girls and 5 boys, **Fig.** 7-3) to often stop in and visit us. This was one of the most joyful times for Mary. We certainly were fortunate to have all of our families near enough to spend holidays (Christmas, 4th of July, etc.) together, because often, due to employment, it is necessary for some members of a family to move a long distance away. In fact, at age 65, I was offered a very attractive Welch Professorship by Texas Christian University, but Mary and I decided not to accept it. This offered an increase in salary, research funding, and a position as long as I remained able to continue my research. The

President at NU, Arnie Weber, did what was possible, but although this was helpful, our decision to stay at NU was made in order to remain near our children.

The other most satisfying, pleasurable thing for Mary was her foursome of bridge each week. The bridge was much less important than was her visits with her good friends Dorothy Letsinger, Faye Peterson, Louise Brittain, and, when needing a substitute, Florence Bowers. The bridge was preceded by lunch and considerable conversation, and Mary would, at times, come home with news of interest to me that I otherwise would not have known. This group was the first to recognize the fact that Mary was having difficulty with her memory. She could no longer remember the cards played. This became so bad that she could no longer play, but she continued to attend the bridge meetings each week, in order to be with her friends. She also had to stop driving her car because she would get confused and have a problem finding her way—even to go to the grocery store where she had routinely shopped each day. More about this later.

Prior to her memory failure, Mary often traveled with me, and enjoyed these trips and visits—to be with friends as well as to meet new people. I continued to do

Figure 7-3. My eleven grandchildren.

considerable traveling in the US and abroad to give invited lectures at universities and scientific conferences. Mentioned earlier was my Humboldt Senior US Scientist Award, arranged by Professor Dr. Edward Schlag, when we spent three months in Munich. During that stay, Mary and I were invited to dinner at the homes of several university professors. We also traveled to several of the major universities in Germany, where I gave invited lectures. One of our trips was to Copenhagen to give a talk at the memorial symposium in honor of Professor Jannik Bjerrum. This was one of the high points of our three months in Europe, because we were able to visit most of the places we liked best during our sabbatical at the University of Copenhagen. Particularly enjoyable was our return to the house, street, and neighbors of that sabbatical year in Denmark (**Fig. 7-4**). Many of the neighbors were still there, and we all enjoyed reminiscing about our friendly year together. During our two days in Copenhagen, our hosts were Claus Schäffer and his wife, who made certain we saw all the places that interested us.

Figure 7-4. My wife Mary and me in 1995, when we were in Copenhagen and visited the street (Kvaedeve) on which we lived in 1954-55, when on a sabbatical in Denmark.

Mary also often accompanied me when I was to receive an Award (see Honors in Appendix). In 1988, we went to Italy, where I was to be awarded the "Laurea Honoris Causa" (Honorary Degree) of the University of Turin. This was particularly significant to me because the university is in the Piedmont region from which my parents came. On that same occasion, I received the Italian Chemical Society Award for Research in Inorganic Chemistry, and the IX Century Medal of Bologna University, founded in 1088. It is the oldest university in the western world, and, some scholars say, in the world. Mary and I enjoyed the receptions and dinners that are part of the celebrations of these Awards. Since we had seen much of Italy during my sabbatical, we decided to spend our remaining days with relatives, Savina Sgarbi and family in Milan (**Figs. 5-6, 7-2**), and Luisa Del Signore and family in Cuneo. In the US, after 1987 when we moved in our new house, Mary attended my reception of the ACS awards—George Pimentel Award in Chemical Education in San Francisco and the Josiah Willard Gibbs Medal Award (**Fig. 7-5**) for chemical research which I received in Chicago. This award is given each year and 26 of the awardees are Nobel Prize laureates, such as Marie Curie and Linus Pauling, etc.

Unfortunately, the plans Mary and I had for after I became Emeritus Professor in our new house lasted only a very short time—four years for Mary and six years for me. In 1991, Mary had an emergency quadruple heart bypass, which was a success, but she immediately went into a coma that lasted a month. A few days after coming out of her coma, the doctor said it would be OK for her to go home. She remained bed-ridden for some weeks; I kept an eye on her during the day, and had a lady come to do so at night. Finally, Mary seemed to have recovered completely, but later some artery blockage occurred that required an angioplasty, for the doctor no longer believed that she could survive surgery. This occurred again when we were in Munich, where they did her second angioplasty. As a result, as soon as Mary was able to travel, she returned home, insisting that I stay to complete my final month of the Humboldt Award.

Soon after her return from Germany, Mary began to show signs of a mental problem, first at her weekly bridge club, then to our family and neighbors. I took her to our doctor and friend, Bernie Adelson, and explained her situation. He said that it appeared to be the onset of Alzheimer's Disease, but that there was no specific test for its diagnosis. He asked Mary to subtract nine from one hundred and she was not able to do it. He told her the brain, like weak muscles, may be helped by exercise. He suggested that we get a book of crossword puzzles which started with easy ones and proceeded to the more difficult. Mary could not even work the first and easiest puzzle. She then began to get lost during her short walks in the neighborhood. The situation became worse each day, and our family got together to talk about how best to take care of her. Fortunately, our neighbor, who is a very good Catholic, suggested we ask about this at the church because they operated some day-care centers to take care of such older

Figure 7-5. Photo taken when I received the Willard Gibbs Medal for "Outstanding Research in Chemistry" in 1996. Front row: me and Mary, then Elizabeth and Fred, Jr. Back row: at left is Margaret and at right is Mary Catherine.

Figure 7-6. Photo taken when I received the Priestley Award in 2001 (see pg. 132) for "Distinguished Services to Chemistry." My children are behind me (left to right): Elizabeth, Margaret, Mary Catherine, and Fred, Jr.

people. We asked about it and were able to arrange this, so I would drop Mary off at the center in the morning on the way to my office at NU, and pick her up later, on my way home.

This worked well because the center only accepted a dozen patients and had two experienced nurses to look after them. Then came the tragic accident that Mary did not survive. About 9:00 AM, I was driving her to the day care center and was only two blocks away when I fell asleep at the wheel. We side-swiped an on-coming truck and I lost control of the car which left the road and hit a tree. The car was totaled, and we were taken by ambulance to the hospital emergency quarters. A quick look at the two of us told the doctor that my condition was not serious, but that Mary's condition was serious. Further examination established that her spleen had been ruptured, and the bleeding had to be stopped or she would bleed to death. The doctor said this would require immediate surgery, and I warned him that she would not survive surgery. He said that he had looked at her hospital history and knew this, but it was the only chance to save her. I agreed that, rather than let her bleed to death, we should take this last chance of surgery. As soon as the doctor had finished the surgery, he came to examine me and to tell me that Mary seemed to be doing OK. A few hours later he returned to inform me that her heart could not sustain the operation, and we would lose her. I had them telephone MC and Fred so that they could start on their way immediately. MC had a two hour drive and Fred about a three hour drive. MC arrived in time to see her mother alive, but Fred arrived a little too late. She died on February 5, 1997. I am told that perhaps I fell asleep because of the combination of some new medications and tiredness—Mary would often awaken at night, and become quite confused and distressed and it would sometimes take hours to calm her down.

My situation after moving into our new place was what I had wanted it to be. I was able to satisfy my hobby of having a small vegetable garden and was able to play golf twice a week on a near-by public course. However, this was short-lived, because the sore back that I had put up with for years became so very painful that I needed help. My doctor and an excellent surgeon that he recommended said the pain was caused by a slipped disk which, when taken care of by surgery, would alleviate the pain. This was just a few weeks from when I was invited to the XXVII ICCC (1989) in Brisbane, Australia (see pp. 176-177). Since Mary had long wanted to go to Australia, I accepted the invitation and the surgery was done once we returned. This surgery was followed by four more surgeries, with the last one taking 14 hours and requiring three surgeons, each of different specialties. The net result of all of this is that the nerves from my brain to my legs were damaged. Now I am unable to walk without two canes, and in our large technological building, I use a scooter (**Fig. 7-7**). I do not recommend it, but being handicapped does have an advantage when it comes to parking and using a washroom. After my fifth operation, I spent two weeks in a rehab ward in an attempt to use the proper exercise to restore my ability to walk. This was not successful, but as often happens in hospitals, I was exposed to bacterium which resulted in a serious case of pneumonia. During my four and a half months in the hospital, I am told I was near death. The doctors were of the opinion that I would not make it so they contacted each of my four children to immediately come to the hospital. Fortunately, I proved them wrong, and they said that I survived because of my strong heart. Now I also

Figure 7-7. Me on my scooter that is needed to get around in our large Tech building. I can "walk" with two canes, but must make frequent stops to rest because of shortness of breath and pain.

have blood clots which hopefully are being taken care of by the blood thinner I must use. Then there is a minor problem of a ulcer. I suppose that beyond the age of 80 what happens healthwise is generally none too good. It is clearly true that one must make the most of their life before they get old and can no longer do some things they enjoy.

In spite of all our health problems, Mary and I felt we had a good life together, and that the time in our life could not have been better (see Chapter 2). We were a happy family, all living near each other, making it possible to often all be together. We had the opportunity to enjoy and be proud of our four children and eleven grandchildren. We never had a real vacation but we did spend a year in Copenhagen and a year in Rome, this being a good time to visit parts of Europe and to make new friends. When the children were old enough to take care of themselves, Mary often traveled with me. Mostly because of the success of the two books I coauthored and because of our research published in scientific journals, I was known by inorganic chemists worldwide. Therefore, during the peak of my professional career, I had many invitations to give lectures, to participate in conferences, and to receive awards in the US and abroad. Mary often came with me during these trips, and thus visited most of the 41 countries listed in the Appendix. Yes, the two of us are thankful for having enjoyed most of our lives together.

In conclusion, I can truthfully say that I marvel at what has happened to me during my life. Not in my wildest dreams could I have expected when growing up in the little mining village of Coello that, some day, I would become a successful chemist. Had I taken my brother's advice, I would have stayed in Coello where I would have lived an easy, normal life instead of having to travel all over the world and having to do all of what was expected of me. This would have been true, but I would not trade my choice and the lifestyle and profession that have been very kind to me for that of staying in Coello.

I have been blessed with success in the field of inorganic chemistry. I have received most of the honors and awards for research and teaching that I might have hoped for in the profession of chemistry, specifically the field of inorganic chemistry. This success could not have been mine had it not been for my parents, Mary and our children, and, of course, the many helpful members of my professional family. As the end is inevitable, I wish to take this opportunity to thank all of you from the bottom of my heart. In the words of my oldest grandchild, John Michael Kunzer, who inherited these words from his grandmother,

Happiness Always.

APPENDIX

FRED BASOLO

Address: Department of Chemistry, Northwestern University
2145 Sheridan Road
Evanston, IL 60208-3113, USA
e-mail: basolo@chem.northwestern.edu
Fax: 847-467-4667

Born: 11 February, 1920; Coello, Illinois

Degrees: B.Ed., 1940, Southern Illinois Normal
M.S., 1942; Ph.D., 1943, University of Illinois

Positions Held:

1943-46	Research Chemist, Rohm & Haas Chemical Co.
1946-79	Instructor, Assistant Professor, Associate Professor, and Professor of Chemistry, Northwestern University (NU)
1969-72	Chairman of the Chemistry Department, NU
1980-90	Charles E. and Emma H. Morrison Professor of Chemistry, NU
1990-present	Charles E. and Emma H. Morrison Emeritus Professor, NU

**Basolo Award Medal for Outstanding Research
in the Field of Inorganic Chemistry**
(See pg. 214)

HONORS

1954-55	Guggenheim Fellow (University of Copenhagen)
1961-62	Senior NSF Fellow (University of Rome)
1964	ACS Award for Research in Inorganic Chemistry
1969	NATO Distinguished Professor (Tech. Univ. of Munich)
1971	ACS North Regional Sections' Citation of Excellence for research on substitution reactions of metal complexes
1972	Bailar Medal Award (first Bailar Medalist)
1974	Southern Illinois Alumni Achievement Award
1975	ACS Award for Distinguished Service in Inorganic Chemistry
1976	Dwyer Medal Award
1977	Fellow of American Association for the Advancement of Science; Honorary Member of Phi Lambda Upsilon
1979	JSPS Fellow (Japanese Society for the Promotion of Science); Member of the National Academy of Sciences
1980	Morrison Professor, Northwestern University
1981	Honorary Member of the Italian Chemical Society; James Flack Norris Award for Outstanding Achievement in the Teaching of Chemistry; NATO Senior Scientist Fellow, Italy
1982	President-Elect of ACS; Illinois House of Representatives Resolution No. 686 honoring FB as a Resident of Illinois
1982-84	Member Board of Directors, ACS
1983	President of ACS; Oesper Memorial Award; Corresponding Member of the Chemical Society of Peru
1984	Honorary Doctor of Science Degree at Southern Illinois University
1985	Honorary Degree at Lanzhou University (PRC)
1987	Foreign Member of Accademia Nazionale dei Lincei (National Academy of Science of Italy); Laurea Honoris Causa, University of Turin
1988	SCI (Italian Chemical Society) Award for Research in Inorganic Chemistry; IX Century Medal of Bologna University; Honorary Degree at Zhongshan University
1990	Harry and Carol Mosher Award; Illinois Governor James Thompson declared 18 August, 1990, the Fred Basolo Day in the State of Illinois.
1991	Padova University Medal; Distincion Bicentenaria Medal of the University of Los Andes in Merida; Medals from the Universities of South Korea, POSTECH, KIST, and the Inorganic Division of the South Korean Chemical Society; Chinese Chemical Society Medal
1992	ACS George C. Pimentel Award in Chemical Education; Chemical Pioneer Award (American Institute of Chemists); Sigma Xi Monie A. Ferst Medal; Humboldt Senior U.S. Scientist Award; Distinguished Visiting Professor, Ohio State University
1993	Gold Medal Award (American Institute of Chemists)
1996	First Lecturer and Medalist of the Royal Society of Chemistry Joseph Chatt Award in the UK; Josiah Willard Gibbs Medal; Inauguration, Member of the Hall of Fame of the Chemistry Department at SIU
1997	Laurea Honoris Causa, University of Palermo; Sacconi Memorial Lecture
2000	SIU Obelisk Leadership Award
2001	Priestley Medal

Dedications of Books and Journals for Commitment to Teaching, Research and Scholarship in Inorganic Chemistry

1990	R. J. Angelici, *Inorg. Synth.* **28**
	U. Belluco, *Inorg. Chim. Acta* **178**
	J. L. Burmeister, *Coord. Chem. Revs.* **105**
	M. H. Chisholm, *Acc. Chem. Res.* **23**, 419
	J. E. Ellis, *J. Am. Chem. Soc.* **112**, 6022
1991	M. Nicolini, *Lectures in Bioinorganic Chemistry*, Raven Press, New York
	S. Gambarotta, *Com. Inorg. Chem.* **11**, 195
1992	J. S. Miller, *Adv. Mat.* **4**, 498
1995	J. L. Burmeister, *Inorg. Chim. Acta*, **240**, XXIII
2000	H. F. Fried, H. B. Gray, and S. H. Strauss, *Chemtracts, Inorg. Chem.* **13**, No. 2
	Helmut Werner, *Organometallics* **19**, 3109
2001	M. A. Reynolds, I. A. Guzei, and R. J. Angelici, *Organometallics*, **20**, 1071

Memberships

U.S. National Academy of Sciences; American Academy of Arts and Sciences; Italian National Academy of Science, Lincei; American Association for the Advancement of Science; American Chemical Society; American Institute of Chemists; Chemical Society (London); Alpha Chi Sigma; Kappa Delta Phi; Phi Lambda Upsilon; and Sigma Xi.

Publications

More than 380 scientific publications. Co-authored two books: *Mechanisms of Inorganic Reactions* with R. G. Pearson (translated in Chinese, Japanese, Russian, and German) and *Coordination Chemistry* with R. C. Johnson (translated in Italian, Japanese, Polish, Russian, Spanish, Korean, Malaysian, and Chinese). Edited Vol. XVI of *Inorganic Syntheses*; co-edited *Catalysis* with R. L. Burwell, Jr., and two volumes of *Transition Metal Chemistry* with J. F. Bunnett and J. Halpern. R. E. Oesper Awardee Interview, Chemistry Dept., University of Connecticut. 1983; D. F. Shriver, Interview, *Coord. Chem. Rev.* **99**, 01, 1990; J. J. Bohning, *Oral History*, 01 March, Chemical Heritage Foundation, 315 Chestnut St., Philadelphia, PA 19106, 1991.

Other Co-Authors

Adamson, A. W.	Collman, James P.	Lester, Joseph E.	Rheingold, Arnold L.
Bailar, Jr., John C.	Dawson, John H.	Lichtenberger, Dennis L.	Sabat, Michael
Baldwin, Jack E.	Dwyer, Frank P.	Lynch, Thomas J.	Sessler, Jonathan L.
Ballhausen, Carl J.	Eliel, Ernest L.	Marder, Todd B.	Shaw, Bernard L.
Barrow, Gordon M.	Ernst, Richard D.	Meeker, Robert E.	Shi, Quizhen
Bjerrum, Jannik	Halpern, Jack	Munson, Ronald A.	Shriver, Duward F.
Bowman, Robert G.	Heck, Richard F.	Nakamoto, Kazuo	Sulfab, Yousif
Bunnett, Joseph F.	Hoffman, Brian M.	Neumann, Henry M.	Swepston, Paul N.
Burwell, Jr., Robert L.	Ibers, James A.	Parry, Robert W.	Trogler, William C.
Cattalini, Lucio	Illuminati, Gabriello	Pearson, Ralph G.	Turco, Aldo
Chatt, Joseph	Ji, Lian-nian	Poe, Anthony J.	Vlcek, Antonin A.
Chen, Yun-ti	Johnson, Curtis E.	Rausch, Marvin	Wasserman, Edward
Chen, Tian-lang	Leal, Orlando	Renshaw, Suan K.	Wilkins, R. G.
Collins, Terrence J.	Lederer, Michel		

Appendix

Ph. D. Students

Anderson, David L. (1975)
Angelici, Robert J. (1962)
Baddley, Wm. H. (1964)
Baker, Bernard Ray (1960)
Bauer, Robert A. (1973)
Beattie, James K. (1967)
Bergmann, John G. (1955)
Boston, Charles R. (1953)
Brault, Albert T. (1964)
Burmeister, John L. (1964)
Cape, Thomas W. (1979)
Carter, Mark J. (1973)
Castor, Wm. S. Jr. (1950)
Clarkson, Steven (1972)
Cohen, Irwin (1964)
Crumbliss, Alvin L. (1968)
Diemente, Damon L. (1971)
Ellis, Paul E. Jr. (1980)
Goddard, John B. (1968)

Gray, Harry (1961)
Hammaker-Taylor, G. S. (1961)
Henry, Patrick M. (1956)
Johnson, Edward D. (1976)
Johnson, Ronald C. (1961)
Johnson, Susan A. (1961)
Jones, Robert David (1978)
Keeton, David Paul (1970)
Kennedy, John Rory (1987)
Kershner, David L. (1987)
Klabunde, Ulrich (1966)
Kolis, Joseph W. (1983)
Kowaleski, Ruth M. (1986)
Linard, Jack E. (1979)
Lofquist, Marvin J. (1969)
Matoush, William R. (1956)
McDonald, John W. (1971)
Messing, Aubrey F. (1957)
Morris, Donald E. (1967)

Murmann, R. Kent (1953)
Musket, Stewart F. (1959)
Myers, Victor G. (1970)
Palmer, G. Todd (1985)
Palmer, Jay Ward (1960)
Raymond, Kenneth N. (1968)
Reed, James L. (1973)
Rerek, Mark (1984)
Richmond, Thomas G. (1983)
Schmidt, Steven P. (1985)
Seiden, Lester, (1957)
Shen, Jian-Kun (1992)
Sheridan, Peter S. (1970)
Stephen, Keith H. (1964)
Stone, Bobbie D. (1952)
Szymanski, Thomas (1978)
Thorsteinson, E. M. (1966)
Weaver, Thomas R. (1967)
Weidenbaum, Kevin J. (1967)
Wojcicki, Andrew (1960)

Postdoctorates

Banerjea, Debabrata (1956)
Bank, Hyochoon (1990)
Belluco, Umberto (1964)
Breitschaft, Siegfried (1964)
Broomhead, John A. (1963)
Budge, John (1979)
Butler, Ian S. (1966)
Carter, Steven T. (1987)
Chang, Chi-Yen (1953)
Cheong, Minsek (1988)
Day, J. Philip (1968)
Dhar, Sanat K. (1961)
Dickens, John E. (1956)
Dokiya, Masayuki (1969)
Druding, Leonard F. (1961)
Edgington, David N. (1962)
El-Awady, Abbas A. (1974)
Ellison, Herbert R. (1960)
Engelhardt, L. M. (1972)
Freeman, Jeffrey W. (1989)
Funke, Lawrence (1977)
Gafney, Harry D. (1972)

Gao, Yici (1986)
Hallinan, Noel C. (1989)
Hashimoto, Toshiaki (1981)
Hoq, M. Fazlul (1989)
Hove, Emmanuel G. (1979)
Jensen, Arne (1956)
Johnston, Ronald D. (1970)
Jouan, Michel (1969)
Kane-Maguire, Leon (1969)
Lane, Bruce C. (1971)
Lane, Ruth M. (1971)
Lange, Bruce A. (1975)
Maggio, Francesco (1964)
Maples, Peter K. (1970)
Mawby, Roger J. (1964)
Meier, Max A. (1968)
Minkel, Daniel T. (1977)
Monacelli, Fabrizio (1961)
Moore, Peter (1965)
Morelli, Giancarlo (1987)
Morris, Melvin (1959)
Nicolini, Marino (1969)

Panasyuk, Vitaly D. (1964)
Panunzi, Achille (1969)
Pawson, David (1974)
Petering, David H. (1969)
Rillema, D. Paul (1973)
Schmidtke, H.-H. (1959)
Schuster-Woldan, H. (1965)
Shimizu, Makoto (1983)
Studer, Tobias (1967)
Summerville, David (1978)
Tanner, Stephen (1966)
Venkatasubramanian, P. N. (1982)
Waind-Nord, G. M. (1959)
Wang, Francis (1972)
Wawersik, Henning (1966)
Weick, Charles F. (1965)
Weschler, Charles J. (1973)
Wise, Gene (1954)
Yamada, Sho (1961)
Yamamoto, Kenichi (1980)
Yunlu, Kenan (1987)
Zingales, Francesco (1964)

SERVICE TO CHEMISTRY

Editorial Boards. Assistant Editor of the *Journal of the American Chemical Society* (1961-64)✦*Chemical Reviews* (1960-65)✦*Inorganica Chimica Acta* (1967-)✦*Inorganica Chimica Acta Letters* (1977-)✦Editorial Advisory Board of *Advances in Chemistry Series, Accounts in Chemical Research, Bioinorganic Chemistry, Coordination Chemistry Reviews, Inorganic Chemistry, Inorganic Syntheses* (Editor of Vol. XVI), *Inorganic and Nuclear Chemistry Letters, Journal of Inorganic and Nuclear Chemistry, The Journal of Molecular Catalysis, Chemical Reviews*✦Advisory Board of *Chemical and Engineering News*✦Advisory Board of *Who's Who in America*✦Editor of *Comments on Inorganic Chemistry* (1988-)✦Editor-in-Chief of *Chemtracts* (1988-)✦*Launched* Gordon Research Conference with its first talk (1951)✦*Launched* ACS journal *Inorg. Chem.* **1**, 1 (1962).

Offices Held: Chairman of the Division of Inorganic Chemistry of the American Chemical Society (1970)✦Chairman of the NU Chemistry Department (1969-72)✦Chairman-Elect (1975) and Chairman (1976) of the Board of Trustees of the Gordon Research Conference✦Chairman-Elect (1978) and Chairman (1979) of the Chemistry Section of AAAS✦President of Inorganic Syntheses Inc. (1978-81)✦Member of the Board of Directors of the ACS (1982-84)✦Member of the Advisory Board of Center for the History of Chemistry✦President of the ACS (1983)✦Member of the Board on Chemical Sciences and Technology of NRC/NAS (1981-86).

Service on Committees and Panels. Research Grants Panels of Air Force Office of Scientific Research (1965-69), Petroleum Research Fund (1967-69), National Institutes of Health (1969-73), North Atlantic Treaty Organization (1971-77, Chairman, 1976)✦Chemistry Department Evaluation Committees at all the Universities of Ontario Canada, University of New Mexico, Syracuse University, Carnegie-Mellon University, Texas Christian University, Duke, Clarkson, Riverside, Macalester, Rutgers, Wyoming, Maryland, North Carolina at Charlotte, McGill✦Ad hoc Committees for ACS Awards in Inorganic Chemistry, the Fresenius Award, the Honorary Membership to PLU Award, the Willard Gibbs Jury, and the Prize Committee in Chemistry of the Wolf Foundation✦Council and Executive Committee of the ACS Division of Inorganic Chemistry✦Council and Board of Trustees of the Gordon Research Conferences (Chairman, 1976)✦Corporation Member and Board of Directors of Inorganic Syntheses (President 1979-81)✦Executive Committee of the International Conferences on Coordination Chemistry✦Inorganic Nomenclature Committee✦National Committee for the International Union of Pure and Applied Chemistry✦NRC/NAS Board on Chemical Sciences and Technology✦NAS Nominations Committee✦NRC/NAS (Pimentel) Report on Opportunities in Chemistry✦ACS Committees on Publication, Education Council Policy, and Professional Relations✦EPSCoR Committee for the State of Kentucky✦Minority Research Center of Excellence Advisory Committee for Chicago State University.

Appendix 229

Consultant. Have consulted for Allied Chemical and Dye, Celanese, US Air Force Wright-Patterson Labs, Amoco, Abbott, Eastman Kodak, Lever Bros., Baxter Travenol and Ashland Chemical♦Evaluated the chemistry departments of the following universities: New Mexico, Syracuse, Carnegie-Mellon, Texas Christian, Memphis State, Duke, Wayne State, Cincinnati, Kuwait, Lanzhou, and all universities in the Province of Ontario♦PEW Science Program for Undergraduate Education proposal evaluation.

Named Lectures: Riley Lecturer, Notre Dame—Welch Lecturer, University of Texas—Frontiers of Chemistry Lecturer, Case-Western Reserve University—Venable Lecturer, University of North Carolina— W. A. Noyes PLU Lecturer, University of Illinois—Francis Clifford Phillips Lecturer, University of Pittsburgh—Distinguished Visiting Lecturer, University of Iowa—Arthur D. Little Lecturer at MIT—Mellon Lecturer at St. Olaf College—Zuffanti Lecturer at Northeastern University— McGregory Lecturer at Colgate Univeristy—Bailar Lecturer, University of Illinois—Dwyer Lecturer, University of New South Wales, Sydney—Otto Mitchell Smith Lecturer, Oklahoma State University—Krug Lecturer, University of Illinois—Chemist/Scholar Lecturer, Ithaca College—R. P. Scherer Lecturer, University of South Florida—Douglas Greenwood Hill Lecturer, Duke University—E. C. Franklin Lecturer, University of Kansas—Floyd E. Bartell Lecturer at University of Michigan—Theme Inorganic Lecturer, Northeastern Teachers Assoc. —Plenary Lecturer at ICCC in Tokyo and in Calcutta—Plenary Lecturer at FEChem in Prague—Plenary Lecturer at Symposia honoring Sacconi in Florence and Malatesta in Milan. ***One or Two Weeks as Visiting Lecture at:*** University of Califonia (Berkeley)—University of California (Santa Barbara)—Harvard University—Melbourn University—Pennsylvania State University—Monash University—University of New Hampshire—University of California (Santa Barbara).

ACS Lecture Tours—ACS-NSF Visiting Scientist Lecturer—University and foreign Chemical Society Lecture Tours in Italy, West Germany, England, Australia, Japan, People's Republic of China, and Peru—Visiting Professor at Kuwait University, Lanzhou University, Lausanne University, and 1990 P. Chini Lecture, 1991—"Tour de Suisse"—Coochbehar Professorship Lecture at the Indian Association for the Cultivation of Science in Calcutta—The First Silver Jubilee Lecture at Utkal University in Bhubaneswar, India—the Joel W. Broberg Lectures at North Dakota State University—the Probst Lectures at Southern Illinois University at Edwardsville.

THE DREAM TEAM OF NATO WORKSHIP, 1972

E. Antonini, Università di Roma
J. C. Bailar, Jr., University of Illinois
F. Basolo, Northwestern University
E. Bayer, Universität Tübingen
W. Beck, Institut für Anorganische Chemie der Universität
S. J. Benkovic, Pennsylvania State University
J. Bjerrum, H. C. Ørsted Institute
G. C. Bond, Brunel University
M. Boudart, Stanford University
R. C. Bray, University of Sussex
H. H. Brintzinger, Universität Konstanz
R. L. Burwell, Jr., Northwestern University
F. Calderazzo, Università di Pisa
S. Carrá, Università degli Studi Bologna
J. Chatt, University of Sussex
A. Cimino, Università di Roma
J. E. Coleman, Yale University
P. Cossee, Shell Research Ltd.
D. A. Dowden, Imperial Chemicals Industries Ltd.
G. L. Eichhorn, National Institute of Health
D. D. Eley, University of Nottingham
C. Franconi, Università di Venezia
J. J. Fripiat, Laboratoire de Physico-Chemie Minerale
F. G. Gault, Université Louis Pasteur
J. E. Germain, Industrielle de Lyon
R. D. Gillard, University of Kent at Canterbury
M. Graziani, Università di Venezia
M. L. H. Green, University of Oxford
W. K. Hall, Gulf R&D Company Ltd.
J. Halpern, University of Chicago
G. A. Hamilton, Pennsylvania State University
P. Henry, Guelph University
J. A. Ibers, Northwestern University
B. R. James, University of British Columbia
C. R. Jefcoate, University of Wisconsin
K. Jonas, Max-Planck-Institut
C. Kemball, University of Edinburgh
H. Knözinger, Institut der Universität
R. J. Kokes, Johns Hopkins University
H. L. Krauss, Freie Universität Berlin
R. Lontie, Katholieke Universiteit te Leuven
P. M. Maitlis, The University
J. Manassen, Weizmann Institute of Science
A. Marani, Università di Venezia
J. C. Marchon, Stanford University
A. Martell, Texas A&M University
K. Mosbach, Lund Institute of Technology
G. W. Parshall, E.I. du Pont de Nemours & Company
R. C. Pitkethly, B.P. Research Centre
E. K. Pye, University of Pennsylvania
A. Rigo, Università di Venezia
J. J. Rooney, Queen's University
S. J. Teichner, Université Claude Bernard Lyon I, and Institut de Recherche sur la Catalyse
R. Ugo, Istituto di Chimica Generale
V. Ulrich, Universitat des Saarlandes
L. Vaska, Clarkson College of Technology
C. Veeger, Agricultural University
J. J. Villafranca, Pennsylvania State University
H. C. Volger, Koninklijke/Shell Lab.
D. D. Whitehurst, Mobil Oil Corporation

Appendix

Countries Visited

Egypt
Australia
Austria
Belgium
Canada
China
Croatia
Czechoslovakia
Denmark
France
Germany
Greece
Hong Kong
Hungary
Iceland
India
Iraq
Israel
Italy
Japan

Jordan
Kuwait
Mexico
New Zealand
Norway
Peru
Poland
Portugal
Singapore
South Korea
Soviet Union, Russia
Spain
Sweden
Switzerland
Thailand
The Netherlands
Turkey
United Kingdom
Venezuela
Yugoslavia

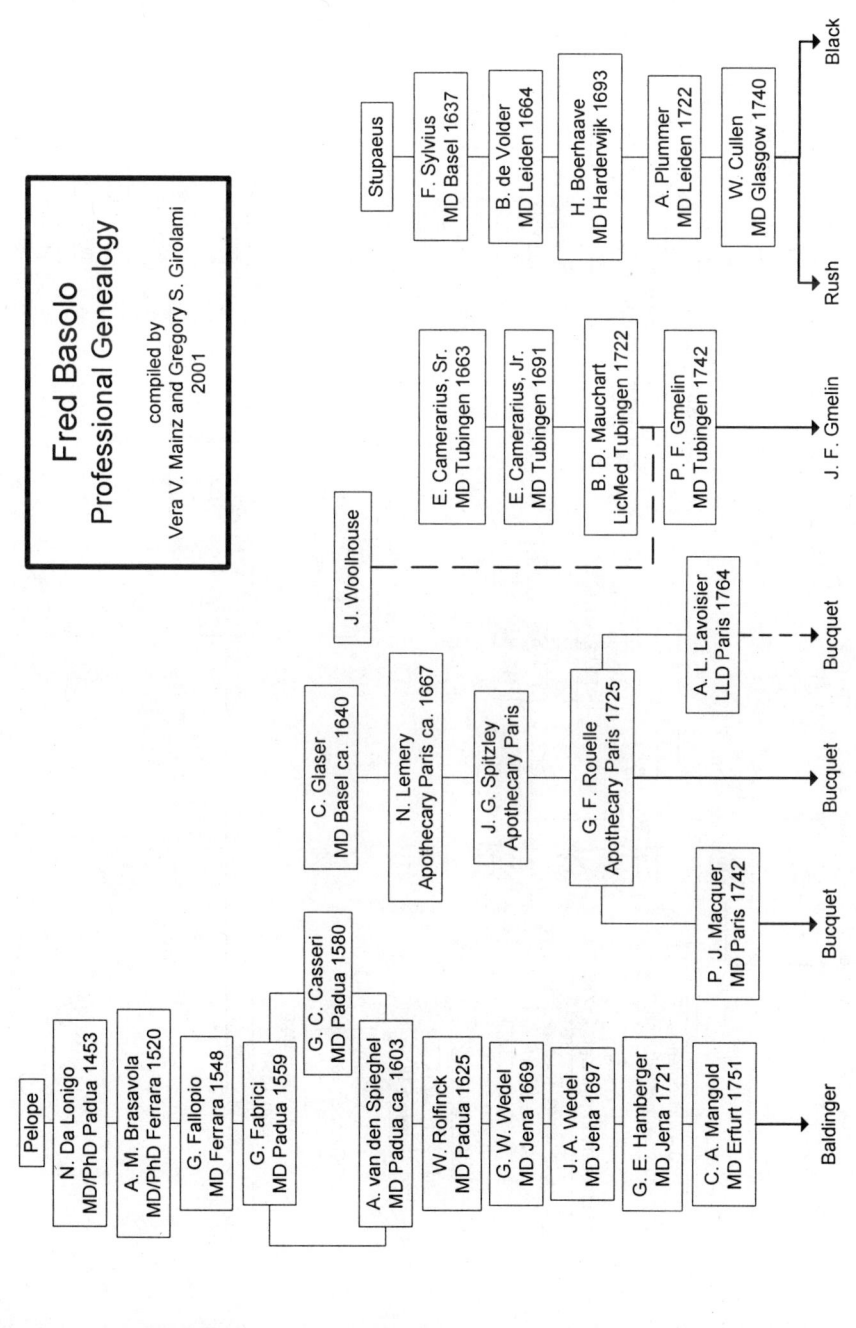

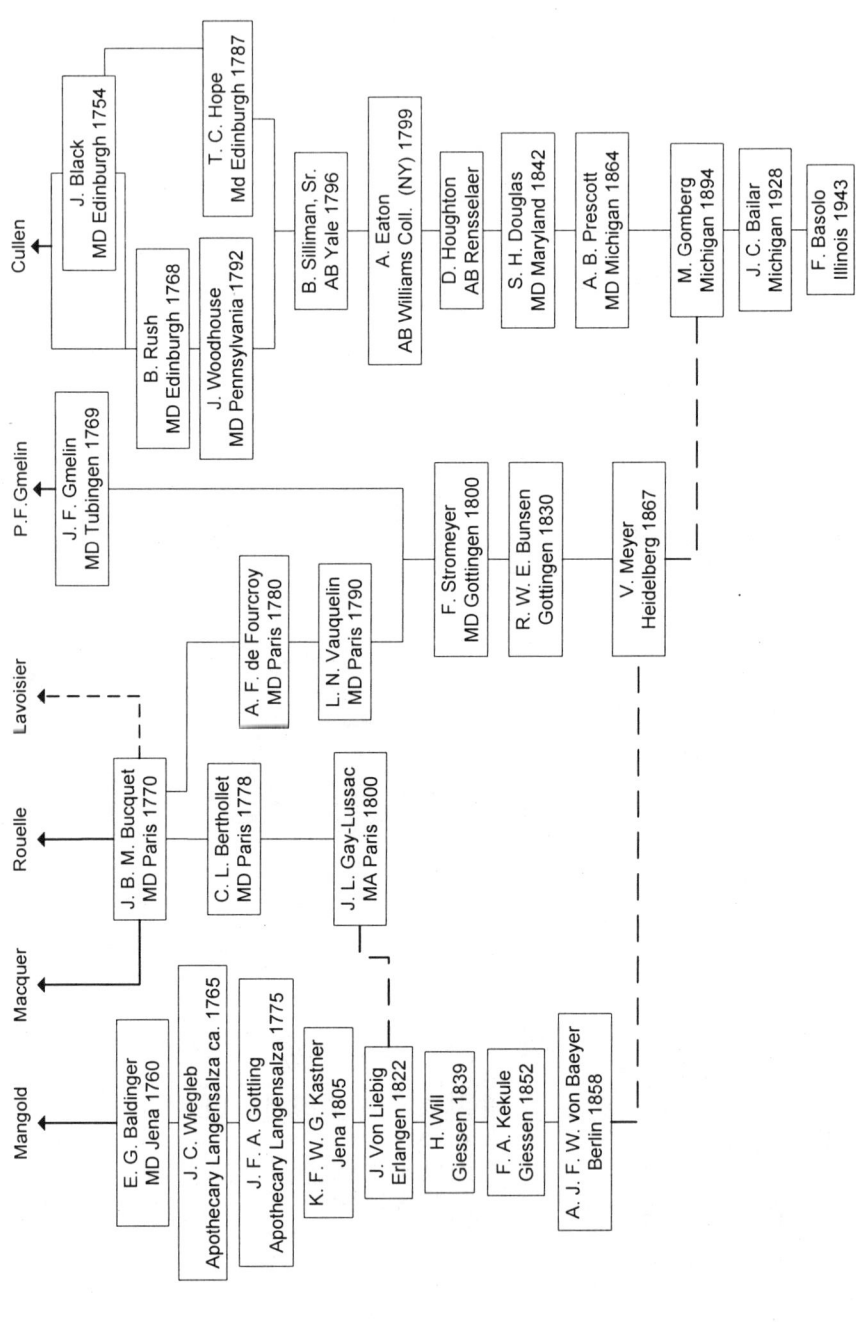

NAME INDEX

Abbott, Talbert, 10, 77
Adams, Roger, 13, 27, 28
Adamson, Arthur, 40, 86, 88, 89
Addison, Clifford, 56, 203, 205
Adelson, Bernard, 35, 177, 218
Al-Obadie, M. S., 177, 179, 180
Al-Shamlan, Ali A., 177
Allen, Professor, 175, 176
Allred, Louis, 79, 116
Anderson, J. S., 57, 75
Andersen, Hans Christian, 39
Angelici, Robert, 81, 103, 104
Asmussen, Prof., 40
Asperger, Smiljko, 188, 189
Audrieth, Prof., 13

Bailar, Jr., John C.,13-15, 17, 18, 21, 26, 28, 76, 80, 84, 93, 127, 133, 139, 144, 206
Bailar, Benjamin, 132
Bailar, Florence, 17, 196, 197
Baker, Robert, 35, 78
Baldwin, Jack, 113, 114
Ballhausen, Carl J., 41
Balzani, Vincenzo, 193
Banerjea, Debabrata, 96
Basolo, Catherina, 47
Basolo, Aldo, 50

Basolo, Anthony, 50
Basolo, Elizabeth (Liz), 33, 61-63, 127, 150, 215, 215, 218, 219
Basolo, Fred, Jr., 36, 38, 39, 43-45, 60-65, 68, 150, 215, 218-220
Basolo grandchildren, 63, 212, 215, 216, 221
Basolo, Margaret (Peg), 60-63, 215, 218, 219
Basolo, Mary, 33-41, 43-65, 68, 119, 129, 130, 154, 160, 161, 163, 164, 166, 171-173, 176, 177, 185, 191, 193, 201, 209, 212, 214-221
Basolo, Mary Catherine (MC), 36, 38, 39, 43, 44, 61, 64, 149, 215, 218-220
Beattie, James K., 176, 183
Beck, Wolfgang, 170, 172
Becke-Goehring, Margot, 191
Belluco, Umberto, 99, 100
Bergman, John G., 88, 91
Bertini, Ivano, 157, 194
Berzelius, J. J., 82, 98
Bjerrum, Jannik, 19, 37, 40, 41, 52, 66, 98, 101, 190, 1`7
Bjerrum, Niels, 40, 76
Bohr, Aage, 45
Bohr, Niels, 40, 45
Booth, H. S., 127

Bordwell, Fredrick, 81
Boston, Charles R., 87
Bowers, Florence, 216
Brault, Albert T., 103
Brintzinger, H. H., 106
Brittain, Louise, 216
Brønsted, J. N., 40, 86
Broomhead, John A., 175, 176
Brosious, Robert, 138
Brown, George, 130
Brown, Theodore L., 107
Buckingham, David, 175, 196
Bunsen, R. W., 191
Burg, Anton B., 139
Burmeister, John L., 93, 94, 207, 214, 215
Burwell, Robert, 36, 71, 78, 151, 152, 191
Busch, Daryle, 43, 82
Bush, George, 181

Cady, George, 139
Caglioti, Vincenzo, 52, 53, 61, 62, 66, 155
Calderazzo, Fausto, 107, 108, 111, 195
Calvin, Melvin, 109, 110
Cattalini, Lucio, 99
Ceccon, Alberto, 195
Chatt, Joseph, 59, 60, 66, 67, 95, 96, 98, 99, 144, 151, 202- 204
Chen, Yun-Ti, 108, 159, 161
Chernyaev, I. I., 24, 94, 95
Coello, Peter, 1
Collman, James P., 114, 214
Columbus, Christopher, 50, 51,
Conant, J. B., 10
Connick, Robert E., 86
Cotton, F. Albert, 105, 141, 171, 181
Cotton, J. D., 176
Cowley, Alan H., 143
Cramer, Robert, 106
Croatto, Ugo, 100
Cruickshank, Alexander, 137, 143
Crumbliss, Alvin L., 81, 110
Curie, M. 142
Curtis, N. F., 174

Dahl, Lawrence F., 214
DeFord, Donald, 189
DeGovani, Maria 47

Del Signore, Luisa, 217
DeVries, Nadine, 140
Dewar, Michael, 59, 95
Dickerson, John E., 57
Diemente, Damon L., 111
Dole, Malcolm, 78, 138
Douglas, Bodie, 139
Drago, Russell, 187, 188
Draper, John, 124
Dunbar, Kim, 140
Dwyer, Frank, 174, 183-187, 196

Einstein, A., 45, 142
Eisenhower, Dwight, 49
Elliot, Dennis, 141, 143
Emeleus, Harry J., 57, 75, 204, 205
Everett, Chris, 150
Eyring, Henry, 65, 66

Fackler, Jr., John P., 143
Fang-Yi, Vice-Premier, 164, 165
Fernelius, W. C., 127, 139
Figgis, B. N., 176
Fischer, E. O., 44, 45, 100, 105, 141, 144, 145, 168, 170, 171, 192
Floriani, Carlo, 111
Fragala, Ignazio, 195
Frost, Arthur, 78
Fuson, Reynald C., 13, 17

Galileo, 158
Gall, John, 139
Garner, Clifford, S., 86
Garrick, F. J., 86, 88
Giorsetti, Maria 46
Godwin, Hilary Arnold, 79
Gomberg, Moses, 14
Gordon, Niel Elbridge, 136, 137
Gore, Albert, 136
Gray, Harry B., 20, 75, 81, 99, 100, 103, 104, 143, 144, 212, 214
Green, Malcom, 214
Grinberg, A. A., 95
Gutman, Viktor, 187, 188

Hadali, Homdallah, 181

Name Index

Haendler, Helmut, 139
Hall, Michael, 113
Halpern, Jack, 214
Hammaker, Geneva, 92
Hanson, Wesley T., 116
Harris, Gordon M., 86
Hartmann, H., 42
Hashimoto, Toshiaki, 118
Hawthorne, M. Frederick, 214
Henry, Pat, 88, 91
Herrmann, Wolfgang A., 172
Heyrovsky, J., 189
Hieber, Walter, 44, 45, 101, 144, 170, 171
Hodel, Donald, 135, 136
Hoffman, Brian, 80, 111, 112
Hoffmann, Roald, 20
Holliday, Bradley, 82
Holm, Richard H., 52, 214
Hopkins, B S, 13, 15, 72
Hoppe, Rudolf, 45
House, D. A., 175
Hume, David, 38
Hupp, Joseph, 80
Hurd, Loran, 29
Hurd, Charles, D., 70
Hussey, Allen, 72

Ibers, James A., 79, 82, 83, 105, 111, 173
Illuminate, Gabriello, 67
Ilse, F. E., 42
Ingold, Christopher K., 80, 84, 87-91, 126, 188, 189
Ipatieff, V. N., 70, 71
Irving, Harry, 58

Jensen, K. A., 40, 44, 60, 144
Ji, Liang-Nian, 107, 167
Johnson, M. D., 208
Johnson, Ronald C., 67, 68, 117
Jørgensen, C. K., 41, 42, 66, 190
Jørgensen, S. M., 18-20, 40, 76, 77, 89, 92

Kauffman, George, 84
Keeton, David P., 206
Kershner, David L., 211
Keyworth, George A., 135, 136
Kida, Shigeo, 119

Kilpatrick, Martin, 139, 211
King, Billy Jean, 150
King, Edward L., 86
King, L. Carroll, 73, 80
Kirschner, Stanley 145
Kissinger, Henry, 165
Klemm, Wilhelm, 45, 46, 66, 192, 193
Klotz, Irving, 78, 80
Kowaleski, Ruth, 81
Krätschmer, Wolfgang, 172

Lamb, A. B., 86
Lambert, Joseph, 80
Lavoisier, Antoine Laurent, 132
Leigh, G. J., 203
Letsinger, Florence, 216
Letsinger, Robert, 72
Lever, Barry, 214
Lewis, Jack, 54, 55, 205
Lindoy, L. F., 176
Lions, Francis, 186, 187
Lunderstrom-Lang, Dr., 89

Madam T, 95, 198
Malatesta, Lamberto, 51, 66, 155, 157, 158, 171
Mao, Chairman, 159, 160, 161, 165, 168
Mariella, Raymond, 72
Marino, Enrico, 50
Marino-Basolo, Catherine, 2
Marks, Tobin, 71, 79, 181, 195, 199
Martell, Arthur, 46, 47, 139
Martin, Jr., S. R., 96
Marvel, Carl, 13, 220
Mawby, Roger, 106
McMillon, E. M., 75
Meeker, Robert, 88
Melchior, Norten, 72
Merrell, Halley, 131
Meyer, Thomas J., 214
Miller, Roscoe, 73
Miolati, Arturo, 19
Mirkin, Chad, 79
Moeller, Therald, 13, 14, 75, 127
Mond, Ludwig, 101
Morris, Donald E., 81, 103, 104

Munson, Ronald, 88
Murmann, Kent R., 92
Murphy, Walter, 71

Natta, Guilio, 155, 156, 197
Nebergall, William, 134, 159
Neckers, James, 9-13
Nesmeyanov, A. N., 198, 199
Neuman, Henry M., 183
Newman, Melvin S., 147
Nguyen, SonBinh, 80
Nicolini, Marino, 99
Nixon, President, 159, 161, 165
Nöth, Hans, 172
Nyholm, Ronald, 58, 59, 66, 84, 88, 90, 91, 126, 186, 192, 193, 205, 206

O'Halloran, Thomas, 79, 194, 212
Oesper, Ralph E., 132
Otsuka, Prof. 118, 173

Parks, George, 137
Parry, Robert, 74, 93, 136, 141, 143
Pasteur, L., 26
Pauling, Linus, 20, 38, 42, 217
Pauson, Peter, 145
Pearson, Ralph G., 42, 72, 78-80, 82-84, 86, 88, 90, 91, 94, 98, 99, 104, 115, 124, 144, 160, 170, 176, 191, 196, 207, 214
Pecile, C., 93
Pedersons, 44
Petering, David H., 112
Peterson, Faye, 216
Pimentel, George C., 122, 123
Pines, Herman, 70, 71
Poë, Anthony J., 207
Poeppelmeier, Kenneth, 79
Press, Frank, 121
Priestley, Joseph, 132, 133

Queenan, Marie, 211

Raymond, Kenneth N., 81, 83, 214
Reagan, Ronald, 10
Reedijk, Jan, 145, 146
Rerek, Mark, 107

Richmond, Thomas G., 108
Robinson, Ward, 111, 112, 175
Romeo, Raffaello, 196
Roosevelt, F. D., 8
Roper, Warren, 174
Rosenberg, Barnett, 25

Sabota, Piotr, 198
Sacconi, Luigi, 51, 52, 66, 154, 155, 157, 158, 194, 202
Sacconi, Maria, 51
Sachtler, Wolfgang, 71
Sagan, Carl, 121
Saito, Kazuo, 117
Salk, Jonas, 147
Sargeson, A. M., 173, 175, 188
Schäffer, Claus E., 41, 190, 217
Schlag, Edward W., 172, 217
Schlesinger, Arthur, 35
Schmidbaur, Hubert, 172
Schmidtke, Hans-Herbert, 88, 89
Schülzenberger, P., 101
Schuster-Woldan, Hans, 105
Schwarzenbach, Gerold, 20, 66, 98, 201
Scott, Robert, 10
Seaborg, Glenn T., 75
Segni, Antonio, 53
Seiden, Lester, 184
Selwood, Pierce, 72, 76, 154, 197
Senoff, Dr., 175
Seyam, Afif M., 181
Seyferth, Dietmer, 100
Sgarbi family, 159, 217
Shaw, Bernard, 99, 144
Shen, Jian-Kun, 211
Shi, Qi-Zhen, 108, 109, 167, 211
Shimizu, Makoto, 119
Shriver, Duward, 79, 181
Shumb, W. C., 139
Sievers, Robert, 146
Sillén, Lars Gunnar, 200
Smalley, Richard E., 172
Smith, Donald, 190
Smith, J. Harold, 137
Stephen, Keith, 67
Stone, Bob, 84
Storm, Carlyle B., 137, 143

Name Index

Strauss, Steven H., 83
Strotz, Robert H., 116
Sulfab, Yousif, 179
Summerbell, Robert K., 73

Tanimoto, Sadao, 118
Tarr, Betty Rapp, 21
Taube, Henry, 20, 86, 87, 90, 92, 115, 141, 148, 214
Thackray, Arnold, 131, 132
Thorsteinson, Erlind M., 81, 103, 104
Tobe, Martin, 90
Trogler, William, 108
Trzebiatowska, Boguslova J., 95, 197
Turco, Aldo, 93, 99

Ugo, Renato, 150, 151
Urey, Harold C., 40

van Lente, Kenneth, 10, 33
Vaska, Lauri, 51
Veillard, A., 112
Venanzi, Luigi, 66, 101, 202
Vlček, A. A., 189, 190
Vol'pin, M. E., 198, 199
Voss, Eric, 212

Walton, Harold, 72, 73
Weatherburn, D. C., 174

Weaver, Thomas, 212
Weber, Arnold, 216 Wells, Vivian, 8
Werner, Alfred, 15, 17-20, 40, 76, 77, 84, 89, 111, 175
Westheimer, Frank, 122
Whitmore, Frank C., 70
Wiberg, Egon, 169
Wilkins, Ralph G., 54, 55, 86
Wilkinson, Geoffrey, 44, 141, 144, 145, 168, 171, 173, 192
Williams, Robert, 58
Williams, Oren, 105
Wojcicki, Andrew, 81, 102-104, 171, 197, 214
Woldbye, Flemming, 41
Wolf, Ricardo Subirana Lobo, 65
Woodward, Robert Burns, 66, 67

Xiao-Ping, Deng, 168

Yamada, Shoichiro, 119
Yanwu-Li, George, 167
Yatsimerski, K. B., 202, 203
You, Xiao-Zeng, 168

Ziegler, K., 156

SUBJECT INDEX

Accademia Nazionale dei Lincei, 157, 158
acid and base hydrolysis of $[Co(NH_3)_5C_i]^{2+}$, 84-90
acid hydrolysis, 86, 87
Air Force Office of Scientific Research (AFOSR), 80, 141
Alpha Chi Sigma, 27
Alzheimer's Disease, 218
American Chemical Society (ACS), 8, 12, 15, 17, 25, 74, 75, 124-136, 146
    accreditation by, 126, 127
    Division of Inorganic Chemistry, 127, 128
    journals, 124, 125
    lecture tours, 125, 126
    president of, 121, 128-132, 135, 136
    representative of, 132-136
Amsterdam, 45, 101, 144
Andre Adora, 61
A. N. Nesmeyanov Institute of Organoelement Compounds, 198
aquation of metal ammine complexes (Cobalt(III) or Co(III)), 86, 87
artificial blood, 113
associative (S_N2) reactions of metal nitrosyl carbonyls, 104-105

Atomic Energy Commission (AEC), 78, 80
Australia, 173-177

B-1 bomber, 110f
Bailar lecture, 17-21
barrier reef, 176
Basolo logo, 212
Basolo Medal and Lecture, 214
Basolo's Groceries, 210
base hydrolysis, 126
base hydrolysis of metal ammine complexes (S_N1CB), 88-91
baseball, 6, 31, 56, 57, 118, 206
BIP, 82, 83, 105, 213
birth of children,
    Elizabeth, 61, 127
    Fred, Jr., 36
    Margaret, 60
    Mary Catherine, 36
birthday (70th) celebration, 212-215
Board on Chemical Sciences and Technology (BCST), 121, 122
boccia, 47
Borgiallo, 3, 49
Boules, 46, 47
bridge, 171, 216

British Open Golf Tournament, 54
brown bag lunches, 41, 151, 186, 190
Brussels, 45, 141, 142, 148, 149, 151

Cambridge, 55-58
Cannes, 46, 47
capped porphyrins, 114
carcinogenic chemicals, 133
Carlesberg mansion, 40
Carlesberg-Tuborg beer, 40, 89
Catalysis, Progress in Research, 152
Catania, 195, 196
chain theory, 76
Chairman of Chemistry Department, 115, 116
Charles D. Hurd Lectures, 116
Chemical Abstracts Index, 124
Chemical Heritage Foundation, 131, 132
chemistry
    description of terms, 21-24
    descriptive, 73, 74
    first course, 7
    major at SIN, 9-11
Chemistry & Engineering News, 125, 129
Chemtracks, 215
Chi Chi Rodriguez Course, 147
Chicago Cubs, 56, 69, 206
China, 159-168
Christopher Community High School (CCHS), 6-8
classified research, 25, 26, 28, 30
Co(III), 84, 86-92
coboglobin, 112
cockney accent, 185, 186
Coello, 7-9, 11, 34, 48, 50, 130
    elementary school, 5, 6
    history, 1
    location, 1
    miners, 3-5
    neighbors, 5
Colfax house, 61, 209-211, 215
Committee on Chemical Sciences (CCS), 121
Communist Party, 65, 190, 198
concorso, 154, 155
Consiglio Nazionale delle Ricerche (CNR), 53, 159

Coordination Chemistry, 67-68, 117, 119
coordination theory, 76
cricket, 56, 57, 206
Croatian Academy of Sciences
Cuneo, 47-49

Danish neighbors, 37-40
dendrimer making use of 22 ruthenium (Ru) metals, 193
Denmark, 19, 20, 217
Department of Energy (DoE), 78, 80, 136, 146
dependence on nucleophilic strength of entering ligand (η_{Pt}), 96-100
dibenzene chromium story, 192
discovery in 1798 of $[Co(NH_3)_6]Cl_3$, 76
dissertation research, 21-26
dissociative (S_N1) reactions of metal carbonyls, 102
Dronero, 2, 48, 49

electron transfer, inner and outer sphere, 87f
ethylenediaminetetraacetate (EDTA), 201, 202
European driving tour, 45-52
European Union (EU), 194
Evanston, 34-36, 60, 193

father, 2-4, 49, 50, 55, 56, 210
    of coordination chemistry in US, 15
    of metal carbonyl chemistry, 44, 101
    of platinum chemistry in UK, 204
    professional, 14
Florence, 145
Four Horsemen, 9
frascati, 64
Frythe, 59
funding, 12, 77-80, 109, 110, 113, 135, 137, 141
    agencies, 146-148

Genova, 50, 51
Germany, 105, 138, 168-173, 217, 218
Glenview, 211, 215

Subject Index

Gmelin's Handbuch der Anorganischen Chemie, 191
GNP, 122, 136
golf, 6, 35, 53-55, 57, 61, 68, 75, 118, 119, 138, 141, 147, 150, 173, 174, 179, 209, 220
Gordon Research Conferences (GRC), 136-143
graduate school (1940-43), 12-14, 16-18, 27, 28
Guggenheim Award, 37

Hainan Island, 167
Happiness Always, 220
Hieber's 80th birthday celebration, 120, 121
honeymoon, 34
honorary degree,
 Lanzhou University, 168
 SIU, 12
 University of Palermo, 158
 University of Turin, 158
 Zhongshan University, 168
housing, 35, 36, 61
How Science Works lecture, 19-21

ICCC, 144-146, 154, 160, 173, 176, 197, 198, 200, 201, 204
Iceland, 148, 149
indenyl kinetic effect, 107
Industrial Associates Program, 115, 116
inorganic chemistry, 29f
 nascent period, 138, 139
Inorganica Chimica Acta, 99, 100, 215
Institute of Organoelement Compounds (INEOS), 199
International Union of Pure and Applied Chemistry (IUPAC), 115, 173
Iraq, 180, 181
Israel, 64, 65
Italian Academy of Science Lincei, 157, 193-195
Italian-French border, 47, 48
Italy, 2-5, 17, 47-52, 59, 61, 153-159, 217

James Flack Norris Award, 74

Japanese Society for the Promotion of Science (JSPS), 117
Journal of the American Chemical Society (JACS), 25, 30, 51, 52, 80, 81, 84, 87, 124
Journal of the University of Kuwait (Science), 181

kinetic *trans*-effect, 24, 99
Kuwait, 177-181
 airplane experience, 177
 life style, 178

Lanzhou University, 109, 167, 168
Laurea Honoris Causa, 158
lecture tour of Switzerland, 20
Leeds, 54
ligand migration reaction, 106
ligand substitution, 125
ligand substitution reactions of Pt(II) square planar complexes, 94-100
linkage isomers, 91-94, 207
 nitro (M–NO$_2$), nitrito (M–ONO), 91-93
 thiocynato (M–SCN), isothiocyanato (M–NCS), 93, 94
London, 58, 59, 150
Long Meadow, 33

mao-tai, 163, 164
marriage, 34, 35
Mechanisms of Inorganic Reactions, 42, 81
mechanisms of ligand substitution reactions
 acid and base hydrolysis of [Co(NH$_3$)$_5$Cl]$^{2+}$, 84-90
 associative (S$_N$2) reactions of metal nitrosyl carbonyls, 104, 105
 dependence on nucleophilic strength of entering ligand (η_{Pt}), 96-100
 dissociative (S$_N$1) reactions of metal carbonyls, 102
 indenyl kinetic effect, 107
 kinetic *trans*-effect, 24, 99
 ligand migration reaction, 106

reaction of $[Co(NH_3)_5Cl]^{2+} + NO_2^-$, 91, 92
reactions of 18-electron versus 17-electron metal complexes, 107-108
ring-slippage associative reactions, 105-107
square planar Pt(II) complexes, 94-100
stereochemical changes of octahedral metal complexes, 85
stereochemical retention of Pt(II) complexes, 98
Milan, 51, 154-156, 159
Ministry of Education, 160-163
mother, 2-6, 46-50

National Academy of Sciences, 15, 63, 117-123, 165
election to, 119
North Atlantic Treaty Organization (NATO), 146, 148-151
Northwestern University (NU), 7, 12, 18, 26, 34, 35, 57, 69, 100, 102, 108, 115, 145, 181, 209, 210, 214
history, 69-72
inorganic chemistry faculty, 79, 80
research 80-84
research environment, 77-80
symposium, 81
Nottingham, 56

Occupational Safety and Health Administration (OSHA), 134
Opportunities in Chemistry, 122
organometallic chemistry, 100-108

Palermo, 154, 155
paramagnetic NMR, 194
parents, 1-4, 8, 9, 153, 158
Petroleum Research Foundation (PRF), 146-148
Ph.D. job market in 1995-97, 212
Piedmont, 2, 47, 158, 217
Pimentel Report, 122, 123
Pinerolo, 50
plexiglass, 28, 32
polemic, 171

Basolo-Pearson/Ingold-Nyholm, 88-91
Werner and Jørgensen, 76, 77
Wilkinson/Fischer, 145
Priestley Medal, 219

reaction of $[Co(NH_3)_5Cl]^{2+} + NO_2^-$, 91-92
reactions of 18-electron versus 17-electron metal complexes, 107-108
recession, 116
relatives, 47-51, 158, 159, 213, 217
Research Cooperation, 77, 80
ring-slippage associative reactions, 105-107
Rohm & Haas, 28-32, 34, 72, 80, 125, 138
classified research, 25, 26
role model(s), 2, 14
Rome, 93, 100, 155
Russia–Workshop–INEOS-94, 199

Sabbatical
Denmark (Professor Bjerrum), 37-44, 217
Rome (Professor Caglioti), 61-68
scooter, 220
Senior Humboldt Fellowship, 171-173, 217
Serbian Academy of Sciences, 189
S_N1 and S_N2 mechanisms, 80, 85, 87-90, 95, 96, 98, 103-108, 126, 196
Southern Illinois Normal (SIN), 8-13, 33, 34, 73
Springfield, 33-35
square planar Pt(II) complexes, 94-100
stereochemical changes of octahedral complexes, 85
stereochemical retention of Pt(II) complexes, 98
submarines, 110f
Summer Schools on Coordination Chemistry, Poland, 197
surgery, 219, 220
Switzerland, 20
synthetic caviar, 199f
synthetic oxygen carriers, 109-114
"artificial" blood, 113
"capped" porphyrins, 114
Co(acacen)L, 110-112
Co(salen)L, 110

Subject Index

coboglobin, 112
L_4MnO_2, 112-113
"picket fence" porphyrins, 114

teaching, 7, 8, 12, 69, 72-77, 163, 169
tennis, 6, 61, 68, 141, 150, 151
thesis mentor, 14-15, 17
Tianjin, 164
trans-effect, 95, 99

Ube Industries, 118, 119
University of Illinois (UI), 12-14, 16, 17, 28
University of Turin, 158, 217
US representative,
    ICCC, 145
    NATO, 148

valence bond theory, 86
Vaska compound, 51
Vatican Academy, 66
Vietnam War, 116, 181f
Vitamin B12, 66, 67

Weizmann Institute, 64
Westman Island, 149
Willard Gibbs Medal, 217
Wimbledon, 150
World Bank, 166f
World War II, 38, 50, 51, 197

Zhongshan University, 167, 168